The Rhizosphere

The Rhizosphere
An Ecological Perspective

Edited by

Zoe G. Cardon
University of Connecticut, Storrs

and

Julie L. Whitbeck
University of New Orleans, New Orleans

AMSTERDAM • BOSTON • HEIDELBERG • LONDON
NEW YORK • OXFORD • PARIS • SAN DIEGO
SAN FRANCISCO • SINGAPORE • SYDNEY • TOKYO

Academic Press is an imprint of Elsevier

Elsevier Academic Press
30 Corporate Drive, Suite 400, Burlington, MA 01803, USA
525 B Street, Suite 1900, San Diego, California 92101-4495, USA
84 Theobald's Road, London WC1X 8RR, UK

This book is printed on acid-free paper. ∞

Library of Congress Cataloging-in-Publication Data
The rhizosphere: an ecological perspective/editors, Zoe G. Cardon,
Julie L. Whitbeck.
    p. cm.
    Includes index.
    ISBN-13: 978-0-12-088775-0 (hardcover: alk. paper)
    ISBN-10: 0-12-088775-4 (hardcover: alk. paper) 1.   Rhizosphere.
2.   Soil ecology.   I. Cardon, Zoe G. II. Whitbeck, Julie Lynn.
    QK644.R445 2007
    577.5′7—dc22

                                                                              2006101249

British Library Cataloguing in Publication Data
A catalogue record for this book is available from the British Library

ISBN 13:   978-0-12-088775-0
ISBN 10:   0-12-088775-4

For all information on all Elsevier Academic Press publications
visit our Web site at www.books.elsevier.com

Printed and bound in the United Kingdom
Transferred to Digital Printing, 2011

Working together to grow
libraries in developing countries

www.elsevier.com | www.bookaid.org | www.sabre.org

ELSEVIER   BOOK AID International   Sabre Foundation

CONTENTS

LIST OF CONTRIBUTORS

Michael Bonkowski, Department of Ecology, Evolution, and Organismal Biology, Iowa State University, Bessey Hall, Ames, Iowa 50011, USA

Andrew J. Burton, Ecosystem Science Center, School of Forest Resources and Environmental Science, Michigan Technological University, 1400 Townsend Drive, Houghton, MI 49931

Zoe G. Cardon, Department of Ecology and Evolutionary Biology, 75 North Eagleville Road, U-3043, Storrs, CT 06269-3043

Weixin Cheng, Department of Environmental Studies, University of California, Santa Cruz, CA 95064, USA

Søren Christensen, Copenhagen University, Biological Institute, Terrestrial Ecology, O. Farimagsgade 2D, DK-1353 København, Denmark

Kristen M. DeAngelis, Department of Plant and Microbial Biology, Koshland Hall, University of California, Berkeley, CA 94720

Peter C. de Ruiter, Environmental Sciences Department, Utrecht University, PO BOX 8011, 53508 TC, Utrecht, NL

Laurie E. Drinkwater, Department of Horticulture, Plant Science Building, Cornell University, Ithaca, NY 14853

Ryan Fimmen, Department of Geological Sciences, Ohio State University, Columbus, Ohio 43210

Mary K. Firestone, Division of Ecosystem Sciences, Department of Environmental Science, Policy and Management, 147 Hilgard Hall, University of California, Berkeley, CA 94720

Catherine A. Gehring, Department of Biological Sciences, Northern Arizona University, BOX 5640, Flagstaff, AZ 86011-5640

Alexander Gershenson, Department of Environmental Studies, University of California, Santa Cruz, CA 95064, USA

Bryan S. Griffiths, The Scottish Crop Research Institute, Invergowrie, Dundee, DD2 5DA, Scotland, UK

Christine V. Hawkes, Section of Integrative Biology, University of Texas at Austin, 1 University Station, C0930, Austin TX 78712

Jason Jackson, University Program in Ecology, Nicholas School of the Environment and Earth Sciences, Duke University, Durham, NC 27708

Nancy C. Johnson, Center for Environmental Sciences and Education, Department of Biological Sciences, Northern Arizona University, Box 5694, Flagstaff, AZ 86011-5694

Noah J. Karberg, USDA Forest Service, North Central Research Station, 410 MacInnes Drive, Houghton, MI 49931

John S. King, Department of Forestry and Environmental Resources, North Carolina State University, Raleigh, NC 27695

Wendy M. Loya, Earth System Science and Policy, Northern Great Plains Center for People and the Environment, University of North Dakota, Box 9011, Grand Forks, ND 58202

Kevin McCann, Department of Zoology, University of Guelph, Axelrod Building, 274, Guelph, Ontario, Canada N1G 2W1

John C. Moore, Natural Resource Ecology Laboratory, Colorado State University, Ft. Collins, CO 80523

Neung-Hwan Oh, School of Forestry and Environmental Sciences, Yale University, New Haven, CT 06520

Kurt S. Pregitzer, Ecosystem Science Center, School of Forest Resources and Environmental Science, Michigan Technological University, 1400 Townsend Drive, Houghton, MI 49931

Daniel deB. Richter, University Program in Ecology, Nicholas School of the Environment and Earth Sciences, Duke University, Durham, NC 27708

Sieglinde S. Snapp, Department of Crop and Soil Sciences, 440A Plant and Soil Sciences Building, Michigan State University, East Lansing, MI 48824-1325

Julie L. Whitbeck, Department of Biological Sciences, CRC-200, University of New Orleans, New Orleans, LA 70148

Donald R. Zak, School of Natural Resources and Environment, Department of Ecology and Evolutionary Biology, University of Michigan, Ann Arbor, MI 48109

ACKNOWLEDGEMENTS

Our work on this project was inspired and sustained by many sources. We thank our colleagues in ecological rhizosphere research, worldwide, for fueling our excitement and curiosity by sharing their new discoveries. Interest generated by a symposium we organized for an Ecological Society of America annual meeting, "The Rhizosphere: Top-Down and Bottom-Up Approaches" (Cardon and Whitbeck, 2000), sparked our motivation to develop this book. Besides the enthusiastic audience response to the speakers' presentations, Dr Fakhri Bazzaz and Chuck Crumley (then at Academic Press), in particular, encouraged us to reinvent the symposium forum in text format. We thank all of the contributing authors for considering their work as facets of a whole integrated rhizosphere system and for their reviews of their colleagues' work within the book. Thanks also to the Elsevier editorial staff – Kelly Sonnack, Julie Ochs, and Meg Day – for their guidance and assistance, and to Kalpalathika Rajan and her team of copy editors, who accomplished the detailed proofreading to polish the rough edges of the text with remarkable speed and professionalism.

REFERENCES

Cardon, Z.G. and J. Whitbeck. 2000. http://abstracts.co.allenpress.com/pweb/esa2000/sessions/109.html, accessed 11/06/06.

Introduction

Julie L. Whitbeck and Zoe G. Cardon

Below the soil surface, the rhizosphere is the crossroads of the soil habitat, a hub of biological, chemical, and physical activity surrounding the living infrastructure of plant roots. Complex fine-scale gradients of substrate availability, water potential, and redox state distinguish this habitat from bulk soil and constrain the distribution and the activity of the tremendously diverse rhizosphere biota. Populations of archea, bacteria, protists, fungi, and animals live here along with plant roots, the activities of each influencing those of the others across spatial and temporal scales spanning orders of magnitude. The nature of the exchange and transformation of universal biological currencies – resources such as organic carbon, mineral nutrients and water – by these biota determines paths of energy flow and shapes community structure and ecosystem properties. Information is also exchanged among rhizosphere inhabitants, via mechanisms including quorum sensing and the production of phytohormone mimics. The influence of rhizosphere activity extends far beyond the rhizosphere itself, manifest across the landscape and through time in patterns of community structure and ecosystem processes, and in patterns of soil development.

Although understanding of soil biological, physical, and chemical function has lagged behind comprehension of aboveground processes, insight into belowground function is essential for grappling with current environmental challenges in natural and managed terrestrial ecosystems worldwide. Interest in rhizosphere ecology has a long history with roots in agronomy, mycology, plant physiology, and microbiology. Contemporary study of the rhizosphere is, by necessity, interdisciplinary, depending on understanding soil physical and chemical properties, plant biology, and the activity and organization of microbes and soil fauna. Syntheses focusing on the rhizosphere (e.g. Fitter 1985; Box Jr. and Hammond 1990; Lynch 1990) reflect this interdisciplinary nature and also note the applied value of rhizosphere research for natural and agricultural ecosystem management. Several more recent offerings have delved into specialized areas of rhizosphere biology, often employing reductionist approaches to examine the nature and function of specific kinds of interactions (e.g. rhizosphere biochemistry in Pinto *et al.* 2001, solute transport in Tinker and Nye 2000, biogeochemistry of trace elements in Gobran *et al.* 2001, and Huang and Gobran 2005), or addressing rhizosphere management in the context of particular goals (e.g. Wright and Zobel 2005).

Within the field of ecology, attention to the rhizosphere has grown extensively and rapidly since the mid-1980s, spanning the full breadth of biological and biogeochemical inquiry from ephemeral shifts in bacterial enzyme production in microliter volumes of soil to landscape scale dynamics of soil genesis over millenia. This book models its cross-scale and interdisciplinary approach after Fitter's 1985 edited volume *Ecological Interactions in Soil*, updating our ecological frame of reference for the rhizosphere. Our goal is to invigorate interaction among scientists working on diverse aspects of rhizosphere ecology and to pique the interest of a broad audience interested generally in belowground ecological function in terrestrial ecosystems. Contributions from a range of scientists focus on rhizosphere ecology and the emergent consequences of rhizosphere activity, including chapters addressing soil biota (plant roots, microbes, soil fauna), interactions among organisms and soils, and the implications of those interactions for rhizosphere trophic organization, productivity, nutrient cycling, soil genesis, and ecosystem management. Instead of writing reviews, authors present perspectives on the rhizosphere, including key developments and the interdisciplinary or cross-scale connections linking their focal research area into the network of ecological rhizosphere research. Complementary views of the rhizosphere from very different spatial and temporal perspectives, and at varied levels of abstraction, overlap in at least three major areas detailed below: rhizosphere soil biogeochemistry and physical structure, taxonomic and functional diversity of rhizosphere biota, and integration and coordination of rhizosphere interactions.

First, a better understanding of the biogeochemical and physical nature of the rhizosphere environment and habitat and a corresponding insight into the influence of rhizosphere ecology on the trajectory of soil genesis emerge from several chapters. Richter *et al.* (Chapter 8) highlight the central role played by roots and associated rhizospheres in chemical and physical weathering, positing that over pedogenic timescales, rhizospheres are fundamental drivers of dramatic soil biogeochemical transformation and overall soil development, so fundamental that almost all soil might be viewed as rhizosphere soil of varying age. Hawkes *et al.* (Chapter 1) begin their chapter with the same notion, then delve into the small-scale yet dramatic redox (Richter *et al.*), resource (Cheng and Gershenson, Chapter 2), water, and other gradients surrounding single plant roots that directly affect microbial community functions such as nitrogen cycling. Such small-scale gradients around roots are illustrated, for example, by rhizosphere-induced soil mottling (Richter *et al.*), and their presence has tremendous implications for ecosystem responses to global change (Cheng and Gershenson, Chapter 2; Pregitzer *et al.*, Chapter 7) when increased [CO_2] and N-deposition induce shifts in fine root turnover, extent of rooting, root respiration, and associated rhizosphere microbial community composition and activity.

Soil physical structure is also influenced strongly by rhizosphere biota and their activities, again with implications from microbial to ecosystem scales. Soil aggregates essential for maintaining soil porosity and conductivity, housing microsites for denitrification, and protecting soil carbon from decomposition are bound and stabilized by fungal hyphae and bacterial exopolysaccharides (Drinkwater and Snapp, Chapter 6; Johnson and Gehring, Chapter 4). Plant roots can lift and mix surface soil layers over generations; great increases in rhizosphere bulk density (and associated decreases in conductivity and pore space) can be driven by single roots as they expand in diameter in deeper soil horizons (Richter *et al.*, Chapter 8). Even the slow, physical breakdown of rock is facilitated by generations of roots as they penetrate channels and generate fractures (Richter *et al.*). The physical structure and organization of rhizospheres themselves, and their persistent imprints on soil physical structure, are emerging as potential key controllers of biogeochemistry and biological interaction across spatial and temporal scales (e.g. Crawford *et al.* 2005).

A second theme is the consideration of patterns and consequences of taxonomic and functional diversity in the rhizosphere. For example, even at global scales, functional biodiversity aligned with evolutionary lineage among mycorrhizae has great implications for large, biogeographic patterns in belowground function (Johnson and Gehring, Chapter 4); saprophytic capabilities are minimal among arbuscular mycorrhizae and maximal among ericoid mycorrhizae, suggesting a biogeographic gradient from grasslands to tundra in the reliance of the mycorrhizal symbiotic partners on free-living saprotrophs for mineralization of nutrients from organic matter. Drinkwater and Snapp (Chapter 6) emphasize the importance of re-establishing biodiversity within the rhizosphere in order to redevelop self-sustaining agricultural systems that require reduced fertilizer inputs. They underscore the idea that, prior to the heavy-input, mechanized agriculture prevalent today, plants and associated rhizosphere biota evolved together within the functioning rhizosphere system, only to be separated conceptually and actually when tillage, fertilizer, and pest control inputs were implemented. These large spatial and long temporal views of patterns in rhizosphere communities are complemented by Hawkes *et al.* and Griffiths *et al.* (Chapters 1 and 3) who focus explicitly on specific community membership and signaling. Communities of rhizosphere microbes and soil fauna shift not only with soil management techniques but also as a function of plant species and soil type (Garbeva *et al.* 2004; Griffiths *et al.*, Chapter 3; Hawkes *et al.*, Chapter 1), yet the implications of diversity for resilience of the soil community and for maintenance of ecosystem functions remain unknown.

The third theme is the quest to understand controls over integrative balance and versatile coordination in the rhizosphere across scales of biological organization. For example, Moore *et al.* (Chapter 5) suggest that the fundamental structure of rhizosphere food webs, with multiple (bacterial, fungal, and root) channels through which energy can flow to higher trophic levels, supports web stability even in the face of shifting food-web membership on ecological, and evolutionary, timescales. It is general community structure, not the specific community members, that is most important in capturing the essence of rhizosphere community energetic function and stability. Hawkes *et al.* (Chapter 1) and Griffiths *et al.* (Chapter 3) suggest that, beyond such generalized trophic relationships, specific chemical communication among diverse rhizosphere community members is a key determinant of ecological function. Notable mechanisms include a plethora of signaling molecules that enable communication and activity coordination among microbes themselves (e.g. quorum sensing signals) and among microbes and plant roots (e.g. molecules similar to plant hormones produced by microbes or by microfauna). Again, in Chapter 6, Drinkwater and Snapp address implications of this flexibility in community composition for agroecosystem management, while in Chapter 7, Pregitzer *et al.* query how resilient rhizosphere community structure will be to global scale shifts in carbon and nitrogen availability.

These contrasting and complementary views suggest that the most powerful insights into the essence of rhizosphere ecology will grow from the synergy of reductionist and integrative ecosystems approaches. Since the organismal diversity in the rhizosphere is enormous, and the suite of potential interactions and mechanistic controllers is too large for all to be examined in detail, broader scale properties, biogeochemical or biogeographical setting, or historical background can be used to help guide the focus and interpretation of mechanistic investigations. The potential implications of newly discovered fine-scale rhizosphere patterns and mechanisms can then be more readily considered in work addressing larger-scale or higher-order ecological system function. Focusing on just one of many promising research directions, recent advances illuminating the nature of the rhizosphere habitat at quite fine-scale resolution, along with a growing appreciation of the relevance of rhizosphere ecological activity for long-term soil development and in the service of human needs, provide the basis for designing studies to investigate the kinds, complexity, and extent of the feedbacks between soil physical structure and ecological processes in the rhizosphere. For example, research examining rhizosphere bacterial physiological responses to variation in soil aggregation can be strengthened by understanding the temporal and spatial scales of variation in soil properties and the ecosystem processes to which these organisms contribute. Likewise, ecosystem scale investigations of soil carbon sequestration can draw upon understanding of the key physiological and community level properties that control carbon and energy flow in the rhizosphere, in order to link changes in soil structure with changes in carbon content over time and/or across landscapes.

From our perspectives as rhizosphere ecologists, we hope the contributions to this book inspire research that draws upon cross-scale and interdisciplinary understanding to develop new insights into rhizosphere ecology and management, as well as into ecology as a whole.

REFERENCES

Box Jr., J.E. and L.C. Hammond (eds). 1990. *Rhizosphere Dynamics.* Westview Press, Boulder.

Crawford, J.W., J.A. Harris, K. Ritz, and I.M. Young. 2005. Towards an evolutionary ecology of life in soil. *Trends in Ecology & Evolution* 20:81–87.

Fitter, A. (ed.). 1985. *Ecological Interactions in Soil.* Blackwell, Cambridge.

Garbeva, P., J.A. van Veen, and J.D. van Elsas. 2004. Microbial diversity in soil: selection of microbial populations by plant and soil type and implications for disease suppressiveness. *Annual Review of Phytopathology* 42:243–270.

Gobran, G.R., W.W. Wenzel, and E. Lombi (eds). 2001. *Trace Elements in the Rhizosphere.* CRC Press, Boca Raton.

Huang, P.M. and G.R. Gobran (eds). 2005. *Biogeochemistry of Trace Elements in the Rhizosphere.* Elsevier, Boston, Amsterdam.

Lynch, J.M. (ed.). 1990. *The Rhizosphere.* John Wiley & Sons, Chichester.

Pinto, R., Z. Varanini, and P. Nannipieri (eds). 2001. *The Rhizosphere – Biochemistry and Organic Substances at the Soil–Plant Interface.* Marcel Dekker, Inc., New York.

Tinker, P.B. and P.H. Nye. 2000. *Solute Movement in the Rhizosphere.* Oxford University Press, New York.

Wright, S.F. and R. Zobel (eds). 2005. *Roots and Soil Management – Interactions Between Roots and the Soil.* American Society of Agronomy, Madison.

CHAPTER 1

Root Interactions with Soil Microbial Communities and Processes

Christine V. Hawkes, Kristen M. DeAngelis, and Mary K. Firestone

1.1 INTRODUCTION

A common definition of soil is "the surface layer of earth, supporting plant life" (Webster's). In fact, most of the volume of the upper weathering layer of the earth's crust has been influenced by plant roots at one time or another, and hence by standard definition, most of the soil would be or would have been at some time considered rhizosphere soil. Here we will focus on soil that is in active, current communion with living plant roots. However, the fact that a large proportion of surface soil was directly impacted by plant roots and associated microbes last year, 10 years ago, or 100 years ago provides a potentially valuable context for discussion of soil microbial communities and processes generally (also see Chapter 8).

Rhizosphere soil effectively forms a boundary layer between roots and the surrounding soil. Because roots and soil act as both sources and sinks for a diverse range of compounds, this boundary layer of soil mediates large fluxes of solution and gas-phase nutrient (and non-nutrient) compounds (Belnap *et al.* 2003). From the microbial perspective then, rhizosphere soil is both a crossroads and a marketplace. The physical extent of the active rhizosphere zone is not easily defined, but at any time is expected to extend only a few millimeters from the root surface and to differ based on the process or characteristic of interest.

Plant roots grow into and through an extraordinary array of "indigenous" soil microorganisms. The phylogenetic and functional characteristics of the community that develops in concert with the plant root is thus framed by the

The Rizosphere: An Ecological Perspective

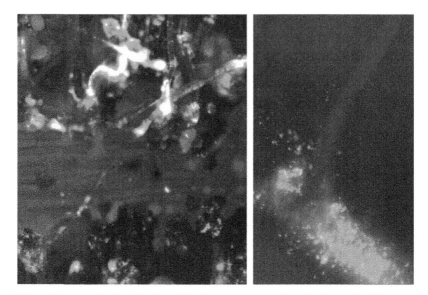

FIGURE 1.1 *Avena barbata* (slender wild oat) roots growing through soil. On the left, magnification is 100×. Plant root and root hairs autofluoresce blue (gray), and soil aggregates infested with bacteria are visible in black at the bottom. Rhizosphere was inoculated with bacteria marked with a constitutively expressing dsRed protein, so all introduced bacteria are visible as red (white) dots. On the right, magnification is 1000×, and bacteria can be seen colonizing the nook between the root and the emerging root hair. Photos by K. DeAngelis. See Plate 1.

background, bulk soil community. While we have been able to photograph relatively intact rhizosphere communities for some time now (Figure 1.1), understanding who these organisms are and what they are doing has been a long-standing challenge. This is an exciting time in rhizosphere microbial ecology. The development of new methods for studying the intact rhizosphere is opening up yet another black box. In this chapter, we discuss recent advances in rhizosphere microbial ecology, the impacts of rhizosphere microbial communities on nutrient cycling, and the importance of rhizosphere processes at larger scales.

1.2   THE COMPOSITION OF RHIZOSPHERE MICROBIAL COMMUNITIES

MICROBIAL POPULATIONS AND COMMUNITIES IN THE RHIZOSPHERE

Plant species can be important in determining the structure of rhizosphere bacterial and fungal communities (e.g., Stephan *et al.* 2000), with both positive and negative effects on different microbial groups. Within plant species, microbial communities can be affected by plant genotype (Smith *et al.* 1999), plant nutrient status (Yang and Crowley 2000), pathogen infection (Yang *et al.* 2001), and mycorrhizal infection (see Chapter 4). Within root systems, microbial communities can even differ among root zones (Yang and Crowley 2000) and at different distances from the root surface as rhizosphere soil grades into bulk soil (Marilley and Aragno 1999). The largest numbers of bacteria in the rhizosphere have been reported to occur in the zone of root elongation (Jaeger *et al.* 1999).

Studying organisms in the rhizosphere, and more generally in soil, is not a straightforward task. A complex community of bacteria may exist at the scale of a soil aggregate, a biofilm, or a section of root surface where boundaries can be difficult to delineate (Belnap *et al.* 2003). Physically removing microbes from soil is also non-trivial, particularly from intact rhizosphere soil. The recent development and popularity of molecular techniques to identify soil organisms has allowed us to move beyond the small subset of culturable soil organisms and begin defining populations and communities of microbes belowground. It is increasingly common to characterize complex microbial communities genotypically using the small subunit 16S ribosomal DNA gene (16S rDNA), a region that is very highly conserved, essential, subject to low homologous gene transfer, and a good reflection of overall phylogenetic relatedness. A collection of 16S genes can be analyzed partially, as with the fingerprinting methods T-RFLP and DGGE, or in detail by sequencing entire populations or communities in clone libraries. Using these methods, we have begun to understand how population and community ecology concepts apply to rhizosphere microbes.

Most population studies have focused on organisms that can be manipulated in agricultural settings either for biocontrol or for increased plant growth, including species of symbiotic nitrogen fixers (Carelli *et al.* 2000), plant growth promoting rhizobacteria (Bevivino *et al.* 1998), deleterious rhizosphere bacteria (Nehl *et al.* 1997), pathogens (Khan and Khan 2002), and bacteriophage (Ashelford *et al.* 2003). Population-level studies are also common for rhizosphere bacteria useful for bioremediation. For example, Dalmastri *et al.* (2003) recently reported high genotypic and phenotypic diversity of a *Burkholderia cepacia* complex population in maize rhizosphere, potentially important in explaining the diverse ecological roles of these bacteria in biocontrol, bioremediation, and human illness.

Because the effects of microbes in the rhizosphere are often synergistic, understanding them at the community level is perhaps most ecologically meaningful. Microbial community characterization is often limited to a subset of the rhizosphere community, such as plant growth promoting bacteria (Dalmastri *et al.* 2003), pseudomonads (Misko and Germida 2002), nitrifiers (Priha *et al.* 1999), or mycorrhizal fungi (see Chapter 4). Alternatively, entire communities can be described. Microbial community characterizations have taken place most often in economically important agricultural species, primarily corn, but also alfalfa, avocado, barley, beet, canola, lettuce, pea, potato, rye, soybean, tomato, and wheat (Table 1.1). In a small number of cases the focus is on plants in natural communities (Priha *et al.* 1999, Kuske *et al.* 2002).

TABLE 1.1   Characterizations of Rhizosphere Microbes Based on 16S rDNA or 16S rRNA

Plant species	Rhizosphere-dominant species	% of clones or bands	Reference
Beta vulgaris	Proteobacteria	50	Schmalenberger and
	CFB group	32	Tebbe 2003a
Brassica napus	Actinomycetes	30	Smalla *et al.* 2001
cv. Licosmos	Proteobacteria (α & γ)	20	
	Gram-positive bacteria	10	
	(*Bacillus megaterium*)		
Brassica napus	α-Proteobacteria	52	Kaiser *et al.* 2001
cv. Westar	(*Bradyrhizobium*)		
	CFB group	30	
	β-Proteobacteria	9	
	γ-Proteobacteria (*Nevskia*)	9	
Dendranthema	Gram-positive bacteria	23	Duineveld *et al.* 2001
grandiflora cv.	(*Bacillus*)		
Majoor Bosshardt	β-Proteobacteria (*Comamonas,*	17	
	Ralstonia, Variovarox)		
	γ-Proteobacteria	17	
	(*Pseudomonas*)		
	α-Proteobacteria (*Acetobacter,*	10	
	Azosporillum)		
Fragaria ananassa	High G+C actinomycetes	50	Smalla *et al.* 2001
	α-Proteobacteria	10	
Hordeum vulgare	γ-Proteobacteria	30	Normander and
cv. Pastoral	(*Acinetobacter, Pantoea*		Prosser 2000
	agglomerans, Pseudomonas)		
	β-Proteobacteria	13	
	(*Burkholderia*)		
	Gram-positive bacteria	13	
	(*Bacillus*)		
Lolium perenne	γ-Proteobacteria	53	Marilley and Aragno
cv. Bastion	(*Pseudomonas*)		1999
	Gram-positive bacteria	15	
	Holophaga-Acidobacterium	15	
	α-Proteobacteria	9	

(Continues)

TABLE 1.1 Characterizations of Rhizosphere Microbes Based on 16S rDNA or 16S rRNA (*Continued*)

Plant species	Rhizosphere-dominant species	% of clones or bands	Reference
Medicago sativa – soil 1	α-Proteobacteria	44	Miethling *et al.* 2003
	γ-Proteobacteria	22	
	Bacteroidetes	22	
Medicago sativa – soil 2	Bacteroidetes	50	
	γ-Proteobacteria	38	
Medicago sativa cv. Regen-SY	Proteobacteria	35	Tesfaye *et al.* 2003
	CFB group	30	
	Gram-positive bacateria	11	
Persea americana	Proteobacteria (*Pseudomonas, Polyangium*)	30	Yang *et al.* 2001
Phaseolus vulgaris	γ-Proteobacteria	60	Miethling *et al.* 2003
	Bacteroidetes	40	
Pinus contorta	α-Proteobacteria	24	Chow *et al.* 2002
	β-Proteobacteria	19	
	Acidobacterium	19	
	γ-Proteobacteria	9	
Solanum tuberosum	Proteobacteria (α & γ)	22	Smalla *et al.* 2001
	Gram-positive bacteria (*Bacillus megaterium*)	11	
Trifolium pratense	γ-Proteobacteria	63	Miethling *et al.* 2003
	β-Proteobacteria	18	
Trifolium repens cv. Milkanova	γ-Proteobacteria (*Pseudomonas*)	52	Marilley and Aragno 1999
	β-Proteobacteria	12	
	Gram-positive bacteria	12	
Zea mays	α-Proteobacteria (Rhizobia)	36	Chelius and Triplett 2001
	β-Proteobacteria (*Burkholderia*)	27	
	γ-Proteobacteria	14	
	CFB group	7	
Zea mays transgenic KX8445	CFB group	24	Schmalenberger and Tebbe 2002
	α-Proteobacteria	21	
	β-Proteobacteria	17	
	γ-, β/γ-Proteobacteria	14	
Zea mays cv. Bosphore and transgenic KX8445	β-Proteobacteria	23	Schmalenberger and Tebbe 2003b
	γ-Proteobacteria	19	
	CFB group	21	
	β/γ- & δ-Proteobacteria	14	
	α-Proteobacteria	9	

In past studies, researchers using culture-based methods have generally reported dominance of Gram-negative bacteria. Results from molecular-based characterizations are, however, more variable, with different groups of dominant microbes in the rhizospheres of individual plant species (Table 1.1). In a meta-analysis of published bacterial 16S rDNA community characterization from rhizospheres of 14 plant species, we discovered that bacteria from rhizosphere soil in fact span the entire tree of life (Figure 1.2). This analysis was based on rhizosphere soils from nine herbaceous dicots, two woody dicots, and three grasses. Bacteria from 35 different taxonomic orders were reported in the rhizosphere. Based on prior results from culture-experiments, we expected to find the Proteobacteria and Actinobacteria well represented, which was indeed the case. Proteobacteria dominated the rhizosphere in 16 of 19 studies (Table 1.1). Within the Proteobacteria, patterns were variable but most often members of the γ-Proteobacteria were dominant. Gram-positive bacteria and the *Cytophaga-Flavobacterium-Bacteroides* (CFB) group followed the Proteobacteria in abundance. Most of the α-Proteobacteria were unclassified at the level of order, which suggests that there is potentially more sequence and functional diversity in the rhizosphere than what was revealed in this analysis. A few unexpected bacteria were found including thermophiles and deinococcus; it is not clear whether these were indigenous soil bacteria or whether these sequences were miscategorized or erroneously sequenced.

Across plant groups, there was a great deal of overlap in the broad taxonomic divisions comprising rhizosphere microbial communities in this analysis (Figure 1.3). The herbaceous dicots exhibited the greatest microbial richness with representatives in 26 orders of bacteria, followed by woody dicots (22 orders) and grasses (20 orders); richness was unrelated to the number of sequences reported for each group. Compositional differences in the rhizosphere microbial community were also evident among the three groups. Relative to the dicot herbs, the woody plants had fewer organisms from the CFB group, Actinobacteria, and Firmicutes, and more Acidobacteria, unclassified β-Proteobacteria, Rhodospirillales, Geobacter, and most orders of α-Proteobacteria. The woody rhizospheres also harbored the only representatives from termite groups and several groups with no cultured representatives including TM6, OP10, and Gemmatimonadetes. Only two woody plant species were included in this analysis, *Persea americana* (avocado) and *Pinus contorta* (lodgepole pine), with the vast majority of sequences contributed by the pine. Pines are well known for their associations with ectomycorrhizal fungi, which may influence the composition of the bacterial community. Very few differences between dicot herbs and grasses could be seen at this coarse taxonomic scale, though they did exhibit slightly different distributions within the Proteobacteria. One study in corn also looked for and found Archaea in the rhizosphere (Chelius and Triplett 2001) and two studies (alfalfa and

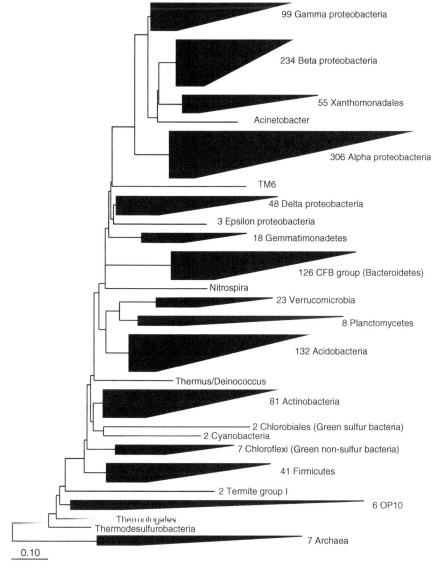

FIGURE 1.2 Evolutionary distance dendrogram of the phyla Bacteria and Archaea based on published 16S rDNA sequences from rhizosphere soils constructed using ARB (Ludwig *et al.* 2004). The dataset consisted of 1227 sequences aligned using an existing 16S rDNA alignment (Hugenholtz, unpublished). In the dendrogram, the horizontal length of each wedge corresponds to the diversity of the group with the scale bar indicating 0.1 changes per site or a 10 percent difference in sequence. Vertical wedge thickness roughly reflects the abundance of different rDNA isolates reported in the dataset; actual numbers of sequences in each monophyletic group are listed adjacent to each wedge.

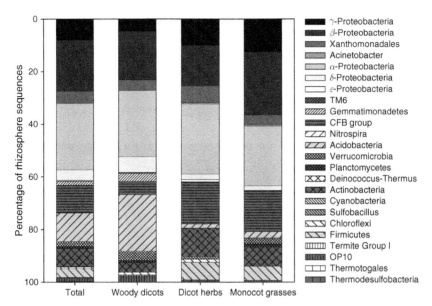

FIGURE 1.3 Stacked bar charts illustrate the relative contributions of bacterial divisions to overall microbial community diversity. The divisions represent monophyletic groups from the phylogenetic analysis and are presented in that order, top to bottom in the legend starting with γ-proteobacteria (see Figure 1.1). On the x-axis, "total" contain all plants in the analysis (see accompanying Table 1.1; n = 1214 16S rDNA sequences); "woody" dicots include *Pinus* and *Perseus* (n = 548); "dicot herbs" include *Beta, Brassica, Dendranthema, Fragaria, Medicago, Phaseolus, Solanum,* and *Trifolium* (n = 376); "monocot grasses" inlcude *Hordeum, Lolium,* and *Zea* (n = 167). All divisions in the legend are represented in the total plants, but may be missing from subdivided plant groups. Relative contributions (%) of each division are based on number of 16S sequences in that division divided by the total number of sequences for each plant group.

corn) reported *Frankia* in the rhizosphere (Chelius and Triplett 2001, Tesfaye *et al.* 2003).

Within herbaceous dicots, differences might be expected in the rhizosphere microbial community of those plants that can and cannot fix nitrogen, given their different requirements for nitrogen from soil. A comparison between the rhizosphere of nitrogen-fixers (*Medicago, Phaseolus,* and *Trifolium*) and other dicot herbs (*Beta, Brassica, Dendranthema, Fragaria,* and *Solanum*) revealed striking differences in the diversity of microbes associated with roots of these two groups (Figure 1.4). Rhizospheres of nitrogen-fixing plants supported a greater richness of bacteria compared to the non-fixers, with nearly double the number of monophyletic groups of bacteria from our analysis. These included the presence of δ- and ε-Proteobacteria, *Nitrosomonas,* Planctomycetes, Deinicoccus-Thermus, Sulfobacillus, and Chloroflexi. Because nitrogen-fixing plants rely less on nitrogen from soil microbial mineralization of organic

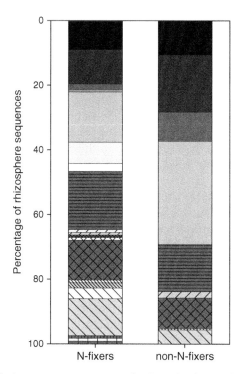

FIGURE 1.4 Microbial community composition of N-fixing (*Medicago, Phaseolus,* and *Trifolium*) and non-fixing dicot herbs (*Beta, Brassica, Dendranthema, Fragaria,* and *Solanum*). For N-fixers, n = 122 16S sequences; for non-N-fixers, n = 254. We did not include records where we could not link 16S sequences to the specific plant (Marilley and Aragano 1999). Legend is the same as for Figure 1.3.

nitrogen, root exudation may be altered and microbial communities may be selected based on characteristics other than rapid growth on labile root carbon.

While this is far from a complete picture of the diversity of the rhizosphere, it demonstrates that the rhizosphere is potentially capable of hosting an array of microbes far more diverse than what has been reported with other methods. It also suggests that, at coarse taxonomic scales, there is some degree of commonality in the bacterial components of rhizosphere communities of many plants, and at the same time there is some degree of specificity in the selection of these communities. As more DNA-based characterizations of rhizosphere microbial communities become available, we can continue to increase our understanding of these communities and their controllers.

Does microbial diversity per se matter in the rhizosphere? Diversity is primarily important in terms of the specific composition of the community present and the amount of functional redundancy included. If multiple pathways for the same process are provided by the community composition, then increased microbial diversity may buffer microbial community structure and function from disturbance (Girvan *et al.* 2005). The rhizosphere is characterized by large environmental fluctuations (see Section "Physical and Chemical Characteristics of Rhizosphere Soil"), which may promote high diversity in the rhizosphere microbial community by maintaining high niche diversity. Thus, microbial community diversity may be important for broad functional continuity in the rhizosphere where disturbances occur in the form of daily environmental fluctuations and in this way has potential for positive feedbacks to the plant (see Section "Plant Populations and Communities").

Community characterization is not always genotypic in nature, but may occur at different scales ranging from functional diversity to broader taxons to simple abundance. Functional diversity is a common measure of microbial community composition and may be more relevant to ecosystem function than taxonomic diversity. Indicators of functional diversity are those that measure the type, abundance, activity, or rate of microbial substrate use. The most common method is the sole-carbon-source utilization profile (Campbell *et al.* 1997). Functional diversity can also be estimated by measuring functional genes that play a role in ecosystem processes (Prosser 2002). This is commonly used for those processes in which a limited number of genes are involved, such as nitrification and denitrification.

In our 16S rDNA analysis of the rhizosphere microbial community, we can relate the phylogenetic diversity to function only through conventional interpretations (Table 1.2). Functional groups relevant to the rhizosphere include nitrogen cycling bacteria, anaerobic bacteria, and pathogens, all of which have been reported in the sequence dataset analyzed here. *Nitrospira*, a genus of microbes important in nitrification, were virtually unreported in these studies. This may have resulted from sampling bias or other methodological constraints, or because this division of nitrifier (which has its ammonia monooxygenase gene unconstrained by internal membranes) is somehow less well suited to the rhizosphere environment. Other common functional groups including methanotrophs and iron oxidizers were also present. Enterics and Xanthomonads (mostly γ-Proteobacteria) represent the majority of known plant pathogens, and these groups were well represented in this sampling.

The phylogenetic and functional composition of rhizosphere microbial communities is the net result of the plant interacting with the indigenous soil community. While we have some knowledge of how the rhizosphere soil environment differs from that of surrounding bulk soil (see below), how the aggregate rhizosphere environmental characteristics select/inhibit specific free-living bacteria and fungi is largely unknown. The specificity of plant–microbial interactions and the environmental characteristics of the rhizosphere are addressed in the following sections.

TABLE 1.2   Functions Traditionally Associated with the Microbial Groups Here Reported in Plant Rhizospheres (Madigan *et al.* 2003). Most of the Listed Functions are not Representative of the Entire Taxonomic Division

Division	Notable members	Characteristics of some members relevant to the rhizosphere
α-Proteobacteria	*Rhizobia, Bradyrhizobia*	methanotrophs, purple non-sulfur bacteria, nitrifiers, diazotrophs
β-Proteobacteria	*Nitrosomonas, Burkholderia*	nitrifiers, diazotrophs
Xanthomonads	*Xanthomonas, Xylella*	common plant pathogens, order of γ-P roteobacteria
Acinetobacter	*Acinetobacter*	twitching motility
γ-Proteobacteria	*Pseudomonas, Vibrio, Erwinia*	methanotrophs, enterics, denitrifiers and some plant pathogens, diazotrophy
δ-Proteobacteria	*Desulfobacteria*	sulfate or sulfur reducers
ε-Proteobacteria	Few cultured reps	?
TM6	No cultured reps	?
Gemmatimonadetes	No cultured reps	?
CFB group (Bacteroidetes)	*Cytophaga, Bacteriodes, Flexibacter, Flavobacter*	capable of growth on complex substrates
Nitrospira	*Nitrospira*	nitrifying bacteria with no internal membranes that house ammonia monooxygenase protein as in other nitrifiers
Verrucomicrobia	Few cultured reps	fermenters (pectin, xylan, starch), oligotrophs
Planctomycetes	*Pirellula*	stalked budding morphology, aerobic chemoorganotrophs
Acidobacterium	Few cultured reps	acid-loving, obligate aerobic organotrophs
Thermus/Deinococcus	Few cultured reps	?
Actinobacteria	Streptomycetes, *Frankia*	high-GC Gram positives, filamentous, prefer alkaline/neutral soils & low water potential, diazotrophs
Chlorobiales	*Chlorobium*	green sulfur bacteria, bacteriochlorophyll, non-motile anoxygenic phototrophs
Cyanobacteria	*Nostoc, Sprilulina*	diazotrophs, filamentous
Chloroflexi	Few cultured reps	green non-sulfur bacteria
Firmicutes	*Bacillus*	low-GC Gram positives, spore-formers, fermenters
Termite group I	Few cultured reps	termite gut bacteria, plant material decomposers
OP10	No cultured reps	?
Thermotogales	Few cultured reps	? (most known members are hyperthermophiles)
Thermodesulfobacteria	Few cultured reps	? (most known members are hyperthermophiles)
Archaea (domain)	Few cultured reps	?

SPECIFICITY OF ROOT–MICROBIAL INTERACTION

Root–microbial interactions encompass a range of specificity from "highly evolved" symbioses (legume-rhizobium) to less-specific associations (arbuscular mycorrhizas). The degree of specificity and coevolution of free-living rhizosphere heterotrophs is, however, quite unclear. As discussed above, host plant genotype can affect the composition of the rhizosphere microbial community and this community can affect plant growth and survival (Nehl et al. 1997). Thus the potential for coevolution exists. Furthermore, colonization of the rhizosphere environment may involve a complex array of microbial behaviors that require the development of some host specificity. A model biological control organism, Pseudomonas fluorescens, was put through the promoter-trapping technique IVET (in vivo expression technology) to determine what genes it needs in order to colonize the sugar beet rhizosphere. Twenty genes were identified as having a significant increase in transcription, of which one quarter were involved in nutrient acquisition (organic acid metabolism and xylanase), three were related to oxidative stress, one was a copper-inducible regulator and one was a component of the type-III secretion system (Rainey 1999).

Apparent symbioses are the most likely to develop host-specificity. Specificity in legume–rhizobia relationships, for example, is determined by plasmids which contain the genes responsible for both nodulation and nitrogen-fixation, and can be exchanged among strains (Hedges and Messens 1990). Some legume–rhizobia associations are clearly defined, with a single rhizobium species limited to a single plant genus or small group of genera (Hedges and Messens 1990, González-Andrés and Ortiz 1999), though individual plants may host several genetic strains of the same microbial species (Carelli et al. 2000) and some rhizobium species are promiscuous. Thrall et al. (2000) posit that rare rhizobia species may be more host-specific than widespread ones.

Non-symbiotic associations may also be specific to plant species or even plant genotype. For example, bacterial antagonists of the soilborne fungal pathogen, Verticillum dahliae, varied significantly among four host plant species (Berg et al. 2002). Plant genotype in hybrid tomato plants accounted for 38 percent of the variability in the plant growth promoting rhizosphere bacterium, Bacillus cereus, mediating resistance to the pathogen, Pythium torulosum (Smith et al. 1999). An examination of replicate plant species in our analysis of rhizosphere microbial communities, however, reveals some repeatable and some unique associations in both Zea mays (corn; Figure 1.5a) and Medicago sativa (alfalfa; Figure 1.5b). In the case of alfalfa and a cultivar of alfalfa, the cultivar had greater microbial diversity with 19 compared to 7 of the monophyletic microbial groups in our analysis represented. Moreover, when alfalfa was grown in two agricultural soils with different management

(a)

(b)

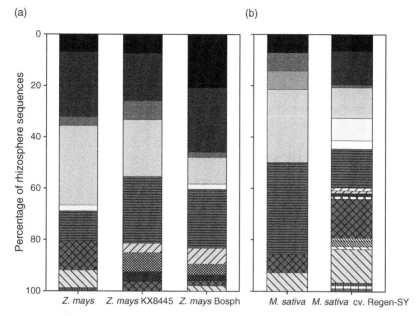

FIGURE 1.5 Microbial community composition in rhizospheres of the same plant species from different studies. (a) *Medicago sativa* (alfalfa) from Meithling *et al.* (2003) (n = 14 16S sequences) is compared to *M. sativa* cv. Regen-SY from Tesfaye *et al.* (2003) (n = 92). (b) *Zea mays* from Chelius and Triplett (2001) (n = 87) is compared to *Z. mays* KX8445 from Schmalenberger and Tebbe (2002) (n = 27) and *Z. mays* cv. Bosphore and KX8445 from Schmalenberger and Tebbe (2003b) (n = 48). Legend is the same as for Figure 1.3.

histories, the alfalfa rhizosphere was dominated in one soil by α-Proteobacteria (particularly rhizobia) and in the other by Bacteroidetes (Miethling *et al.* 2003). Microbial community composition in the rhizosphere of corn and two corn cultivars has a good deal more overlap, with only two microbial groups unique to one cultivar, *Z. mays* Bosphore, which was grown in the same soil as *Z. mays* KX844 but at a different time. Though this small sample is hardly conclusive, it furthers the case that microbial communities in the rhizosphere are primarily non-specific and are selected through a combination of the available bulk soil microbial pool, plant species, and environmental conditions.

1.3 CHARACTERISTICS OF RHIZOSPHERE SOIL THAT IMPACT MICROBIAL COMMUNITY COMPOSITION

The physical, chemical, and biological environment that rhizosphere soil provides for microbial growth can differ substantially from that of nearby bulk soils. The differences in rhizosphere and bulk soil encompass virtually every environmental determinant that is critical to soil microbial activity and survival. Moreover, the biota of the rhizosphere environment differs substantially from that of bulk soils in abundance, composition, and trophic interactions (see Chapters 3, 4, and 5). The compositional and functional characteristics of the rhizosphere microbial community are thus determined by the integrated environmental determinants operating on the bulk soil microbial community together with the biotic interactions occurring in the rhizosphere habitat.

PHYSICAL AND CHEMICAL CHARACTERISTICS OF RHIZOSPHERE SOIL

Microbes in the rhizosphere are subject to an environment in which the supply of water, oxygen, and nutrients is strongly influenced by plant activity. An actively transpiring plant removes huge quantities of water from the soil. Depending in large part on the rate of water supply from the surrounding soil to the rhizosphere, the water potential in rhizosphere soil can be more than 1 MPa lower and much more variable than in the surrounding soil (Papendick and Campbell 1975). During the daytime, rhizosphere soil is commonly measurably drier than the surrounding bulk soil.

In contrast, rhizosphere soil in some terrestrial ecosystems can exhibit higher water content than that of the surrounding soil at night as a result of "hydraulic redistribution" (Caldwell and Richards 1989). Water from deeper in the soil profile is accessed by deep roots, transported to roots in surface soil, and can ultimately move out into dry surface soil at night when evapotranspiration from leaves is reduced. Both of these preceding phenomena can result in large diurnal water potential fluctuations in the soil adjacent to roots, fluctuations that likely are a critical environmental characteristic selecting rhizosphere microbial communities and influencing the rates of nitrogen-cycling occurring in this zone.

The rhizosphere zone is also characterized by high rates of O_2 consumption caused by both root and microbial respiration (Sorensen 1997). This respiration can create zones of low O_2 concentration and even anaerobic conditions depending on the diffusional resupply of O_2 into the rhizosphere from surrounding soil pores. Because diffusion of O_2 is highly dependent on soil water content, reduced water content in rhizosphere soil pores due to plant evapotranspiration can result in enhancement of O_2 diffusion. Thus, depending on the water content of soil and connections to oxygen-depleted or oxygen-replete atmosphere, the availability of O_2 in the rhizosphere soil atmosphere can be either greater or lesser than that of the surrounding soil.

Plant roots are also well known to change the pH of the rhizosphere by extruding protons via H^+-ATPase in epidermal cells (Hinsinger et al. 2003).

This can occur, for example, in response to iron deficiency (Schmidt *et al.* 2003), since a change in pH affected by the plant can also cause the release of inorganic metals. Low molecular weight organic acids secreted by the plant can also act to lower the pH of the surrounding soil.

The rhizosphere is thus a spatially and temporally patchy environment with rapid (commonly diurnal) fluctuations between potentially extreme conditions, including cycles of water stress and anaerobiosis, that microbes must respond to in order to survive and thrive.

RESOURCE AVAILABILITY TO MICROBES IN RHIZOSPHERE SOIL

Plant roots exude a large amount and a complex assortment of organic compounds into the nearby soil (see Chapter 2). The quantity and quality of these carbon inputs vary with plant species, genotype, age, physiological status, root morphology, and the presence of the solid soil matrix, soil organisms, and water availability (Neumann and Römheld 2001). The variety of carbon compounds also changes with location along the root (Jaeger *et al.* 1999). The quality of the carbon, as a substrate and as a chemically active input to the soil environment, is a critical determinant of the composition of the community that results from the root interaction with the extant soil community. In addition to simply acting as a resource, root exudates can influence biotic interactions by attracting beneficial and pathogenic organisms (see Chapter 4) (Nehl *et al.* 1997, Tesfaye *et al.* 2003).

Organic carbon exuded from roots is not the only resource required by rhizosphere microbes. Transpiration-driven movement of water carries nutrient and non-nutrient salts dissolved in the water from bulk soil through rhizosphere soil. This flux of soluble salts into the rhizosphere can result in nutrient and non-nutrient salt concentrations many times that of bulk soil. Conversely, nutrient ion uptake by roots drives diffusional movement and creates zones of nutrient depletion (e.g., NH_4, NO_3, and PO_4) in rhizosphere soil.

BIOTIC INTERACTIONS

A variety of biotic interactions occur in the rhizosphere that can affect the diversity and composition of the microbial community associated with roots. These include interactions of roots with mycorrhizal fungi and root pathogens, root–bacteria interactions, interactions within the microbial community, and interactions of mesofauna with the microbial community. Biotic interactions are typically more common in rhizosphere than in bulk soil, and in many cases may equal or exceed the importance of abiotic factors in determining microbial community composition.

Bacteria and Roots: Chemical Signaling

When a root passes through soil and activates the indigenous microbial community there, competition among microbes for resources or space will partially determine the resulting rhizosphere community. During this process, bacteria may interact with each other, and with roots, through the release and detection of organic signaling molecules. Since the 1990s, interest has particularly developed in quorum sensing in the rhizosphere, in which symbiotic, pathogenic, or plant-associated soil bacteria perceive a threshold concentration of chemical signal (i.e., "sense a quorum"), inducing a change in gene expression and thus behavior (Loh *et al.* 2002, for a brief review). The nature of the chemical signal can be specific to one organism or more general (Loh *et al.* 2002), with more closely related microbial species generally sharing more similar signals. For Gram-negative bacteria these are commonly acyl homoserine lactones; for Gram-positive bacteria, modified polypeptides have been found (Miller and Bassler 2001). Many known soil bacteria have at least one receptor that senses signal and a synthase that makes signal, and some rhizosphere bacteria are known to make a number of signal compounds (Loh *et al.* 2002).

Chemical signaling allows individual bacterial cells to act in coordination with an entire bacterial population or community in response to the presence of new resources or environments such as the rhizosphere provides. Coordination may result in either cooperative or competitive ecologically relevant behaviors in the rhizosphere that would have a lower chance of success if undertaken alone, though Crespi (2001) emphasizes the challenge in demonstrating that a group behavior in microbes is adaptive. Many biotic interactions involve chemical signals in response to environmental cues that induce specific responses including pathogenesis, release of extracellular enzymes, antibiotic production, biofilm formation, symbiosis initiation, and motility (Loh *et al.* 2002).

Direct interactions between the root and the bacterial community may also occur (beyond the signaling long known in the symbiosis literature), particularly when quorum-sensing behavior is triggered by the root or other bacteria. Plants have the ability to affect density-dependent behaviors of rhizosphere bacteria by secreting quorum-sensing mimic compounds or interference compounds (Miller and Bassler 2001). Exudates from pea (*Pisum sativum*), for example, repressed violacein synthesis, extracellular protease activity, and chitinase activity in *Chromobacterium violaceum* and induced swarming in another Gram-negative bacterium, *Serratia liquefaciens* (Teplitski *et al.* 2000). The microbial community has an elaborate and varied repertoire of signaling mechanisms that can affect plants as well. Bacteria can sense and respond to phytohormones from plants and release hormone analogues, resulting in either positive (e.g., plant growth promoting bacteria) or negative (e.g., deleterious rhizobacteria or pathogenic bacteria) effects (Barazani and Friedman 2001).

Mycorrhizal Fungi and Root Fungal Pathogens

When roots are infected with mycorrhizae, rhizosphere microbes are likely affected by changes in carbon exudation brought about because the fungus is a large sink for plant carbon and may therefore impact both the quantity and the quality of carbon leaving the root. The same could be true for any pathogenic fungus that changes root carbon flow. Such interactions are complex and likely species- and host-specific (Schwab *et al.* 1984, Marschner *et al.* 1997, see Chapter 4). Mycorrhizal infection of roots can affect the rhizosphere bacterial community, though the direction of the effect is variable. Some groups have reported no change in microbial activity, number, or composition at coarse taxonomic scales, whereas others have shown that mycorrhizal infection can change the fine-scale taxonomic composition of the rhizosphere bacterial community in a manner that is dependent upon the plant–fungal species pair (see Chapter 4). Some bacteria appear to require the presence of mycorrhizal hyphae (Andrade *et al.* 1998a) and some can be found closely associated with hyphae (Bianciotto *et al.* 2000). Conversely, the presence of certain bacteria can enhance mycorrhizal colonization of roots (Garbaye 1991). Root pathogens compete for space with mycorrhizal fungi and may both directly and indirectly affect the composition of the bacterial community in the rhizosphere (see Chapter 4). Ultimately the interactions among bacteria, mycorrhizal fungi, and root pathogens in the rhizosphere may largely determine microsite occupancy.

Mesofauna and Rhizosphere Microbial Communities

Roots and their associated bacterial communities also attract predators, including protozoa, nematodes, enchytraeids, mites, and collembolans, that can directly influence the abundance and composition of bacteria (Garbaye 1991). These complex and interesting rhizosphere food webs are discussed elsewhere in this book (see Chapters 3, 4, and 5). Clearly, the composition of the rhizosphere microbial community reflects selective grazing by mesofauna in the short term as well as more complex biotic interactions over evolutionary time.

1.4 IMPORTANCE OF RHIZOSPHERE MICROBIAL COMMUNITIES AT LARGER SCALES

Are differences in the rhizosphere microbial community meaningful at a larger scale? A shift in microbial composition can be functionally important such that processes in the rhizosphere may be affected even when overall levels of microbial diversity stay the same. More diverse microbial communities may also be more functionally resilient to disturbance (Girvan *et al.* 2005). Here we briefly consider how rhizosphere microbes can affect plants and soil processes, and how these effects may scale up to influence landscape patterns and ecosystem functioning. We also refer readers to Chapters 3, 4, 7, and 6 for perspectives on the rhizosphere's role influencing nutrient cycling, plant community composition at the landscape scale, global change, and agro-ecosystems, respectively.

Plant Populations and Communities

Microbial communities in the rhizosphere can directly impact plant productivity and demographic parameters. Root pathogens are the obvious case, acting as a sink for plant carbon, damaging root tissue, reducing root uptake, and directly reducing plant growth, reproduction, and survival (Weste and Ashton 1994, Packer and Clay 2003). Microorganisms in the rhizosphere that are more mutualistic can also affect individual plant performance. Arbuscular mycorrhizal (AM) fungi, for example, altered growth and flowering of grasses and forbs in tallgrass prairie, with the direction of the effect dependent on the species (Wilson *et al.* 2001). Free-living microbial communities in the rhizosphere are more likely to have indirect effects on plants. Microbes can function as competitors for scarce nutrients in the rhizosphere (Kaye and Hart 1997), but may also increase availability of resources such as nitrogen to the plant (see below). By changing resource availability to plants, microbes can alter plant fitness and ultimately population dynamics.

Microbes living in the rhizosphere can also affect aboveground plant community composition and, potentially, successional trajectories (Reynolds *et al.* 2003). Simple feedbacks between soil microbial communities and plant communities can alter plant community composition. Plants that experience a net negative effect of soil rhizosphere microbes will ultimately be replaced, whereas those that experience a net positive effect will continue to occupy the patch (Bever *et al.* 1997). Negative feedback can thereby maintain changing diversity through time whereas positive feedback should support a static plant community composition. This model, which assumes that plants develop a unique suite of rhizosphere microbes, fits particularly well for AM fungi, and both positive and negative feedbacks of AM fungal communities on plants have since been demonstrated. Soil pathogens can also drive shifts in the dominance of grassland patches through negative feedback effects (Olff *et al.* 2000). Furthermore, small-scale distance- and density-dependent mortality of conspecifics caused by soil pathogens can be observed. This is the case with black cherry (*Prunus serotina*) seedlings, which are subject to increasing mortality from *Pythium* spp. closer to parent trees (Packer and Clay 2003). In these ways, rhizosphere microbes can drive patch dynamics in plant communities.

Landscape level changes to plant community composition can also be caused by root-associated microbes. Dramatic examples of this can be found with the invasion of root rot pathogens in the genus *Phytophthora* into new habitats. *Phytophthora cinnamoni* invasion of Australian open forests caused the death of >40 percent of the dominant eucalypts (*Eucalyptus* spp.) and the complete destruction of the dominant sclerophyllous understory shrubs (*Xanthorrhoea australis*) (Weste 1986). Thirty years after pathogen invasion, the aboveground community composition has not recovered (Weste and Ashton 1994). In the northwestern United States, a congener, *Phytophthora lateralis*, caused 46 percent mortality of Port Orford cedar (*Chamaecyparis lawsoniana*) populations and 10 percent mortality in nearby Pacific yew shrubs (*Taxus brevifolia*) across Oregon and California (Murray and Hansen 1997). Pathogens in the rhizosphere can have extremely widespread effects on the composition of plant communities, and their presence may in part define the distributional limits of some species.

The distribution of species across the landscape can also be defined by the presence or absence of required symbionts. For example, the ranges of two woody legumes (*Genista* spp.) in Spain are effectively limited by the distributions of specific rhizobia, as they cannot establish in new sites in their absence, but the bacteria are only found in soils where the shrubs already naturally occur (González-Andrés and Ortiz 1999). In agroforestry, pine plantations had little success until they were grown with the simultaneous introduction of associated ectomycorrhizal fungi (Allen 1991).

While the impacts of plant symbiotic and pathogenic microorganisms on plant communities are well documented, the possible effects of free-living bacteria and fungi in rhizosphere soil are again much less apparent. Recent work examining plant-associated microbial N-processing in a California grassland demonstrated plant population effects on the composition of the bacterial nitrifying community and suggested that the resulting impacts on NO_3 availability provided a feedback loop that impacted plant nutrient status, with the potential to influence colonization and competition in this ecosystem (Hawkes *et al.* 2005).

ECOSYSTEM PROCESSES

Root-associated soil microbial communities are in many cases important drivers of ecosystem processes and can control process rates. Many of the plant–microbe interactions that affect plant communities will also affect ecosystem functions by changing primary productivity, plant community composition, plant inputs to soils, and food web structure. Rhizosphere microbes can directly affect ecosystem functions including both nutrient and carbon cycling and storage. We will focus on the role of rhizosphere microbes in the soil nitrogen (N) cycle, which is entirely microbially mediated. Some N-transformations are carried out by diverse groups of microorganisms (N-mineralization) while others are controlled by small, specific groups of microbes (nitrification).

Microbial N-processing in rhizosphere soil is of central importance in the availability of N to plants because it is the location where: (1) N is actually taken up by plant roots; (2) root processes immediately impact and interact with microbes actively transforming N; and (3) head-to-head competition for nitrogen might occur. Rhizosphere microbial communities are known to differ from those in bulk soil in terms of metabolic profiles, activity, and species composition (e.g., Klemedtsson *et al.* 1987, Sorenson 1997, Yang and Crowley 2000) suggesting that rhizosphere N-cycling may be substantially different that that in bulk soil.

Nitrogen Mineralization

The conversion of organic N to NH_4^+ generally occurs under three conditions in soil: when (i) microbes are carbon starved and utilize the keto skeletons of amino acids for energy generation; (ii) fluctuations in water and temperature cause cell death and lysis; the subsequent utilization of the low C/N necromass results in N-mineralization; and (iii) microbes are consumed by predators that release excess NH_4^+.

The increased numbers of microorganisms in rhizosphere soil can represent a potentially labile stock of organic N near plant roots. There are several ways that N contained in microbial biomass can become available to plants. If the supply of labile C is high near young roots and declines substantially in older root sections, then C-limited heterotrophs would mineralize NH_4^+ during catabolism of N-rich cell components. Such a spatial pattern of C-availability along roots (high C availability near root tips and low C availability near mature roots) could in itself result in N-mineralization. Alternatively, root-carbon enhancement of microbial numbers and activity may attract bacterivores, which upon consumption of low C/N microbial biomass release N as NH_4 into the rhizosphere (see Chapter 2). Alternatively, rhizosphere bacteria could be infected by bacteriophage (Ashelford 2003); this would also result in cell lysis and biomass N-mineralization. Finally, rhizosphere soil is a zone of water potential fluctuation as a result of evapotranspiration during the day followed by re-equilibration with surrounding soil water during the night. Such relatively rapid fluctuations in soil water potential could also result in N-mineralization from the rhizosphere microbial biomass as N-rich cellular materials are released during cell water potential equilibration (Kieft *et al.* 1987, Halverson *et al.* 2000).

The potential for N-mineralization in rhizosphere soil thus appears to be high. We have recently measured average gross rates of N-mineralization in rhizosphere soil of *Avena barbata* that were 10 times higher than in bulk soil (Herman *et al.* 2006). While plants may be able to affect rates of N-mineralization by a variety of mechanisms, an intriguing possibility is that microbe–microbe and root–microbe communication can affect N-cycling in the rhizosphere.

The traditional view is that plants obtain virtually all of their N from inorganic sources and compete poorly with soil microbes for NH_4^+ and NO_3^-. This standard view of the N-cycle is, however, undergoing major evolution. Schimel and Bennett (2004) have recently suggested that depolymerization of N-containing macromolecular polymers by soil microbes drives N-cycling in soil. Because the bulk of soil N is organic – primarily chitin, proteins, ligno-proteins, and nucleotides – microbial production of extracellular enzymes that release N in more accessible monomeric forms may be mediating the rate-limiting steps in the production of root-available N (Badalucco *et al.* 1996). Interactions between roots and soil heterotrophs that result in increased activity of enzymes involved in depolymerization of macromolecular organic N is thus highly relevant to root N-availability. Recent work in our lab (DeAngelis, unpublished) has shown elevated activities of N-Acetyl Glucosaminidase (chitinase) and protease in rhizosphere soil adjacent to *Avena barbarta* roots. The activities of these two key enzymes differed in soils adjacent to different root zones. The production of some of these enzymes by Gram-negative bacteria has been found to be under the control of signaling molecules, acylated homoserine lactones (Loh *et al.* 2002). Chemical interactions among bacteria and roots may play a significant role in controlling plant N availability in rhizosphere soil.

Nitrification

Autotrophic nitrification in soil is thought to be primarily limited by the availability of substrate (NH_3/NH_4^+). Historically, plant roots are believed to depress rates of nitrification by three possible mechanisms. First, if rates of root NH_4^+ uptake exceed rates of resupply, then zones of NH_4^+ depletion occur in rhizosphere soil, thus limiting nitrification. Second, roots supply carbon to the rhizosphere; any factor which increases carbon availability potentially enhances net NH_4^+ immobilization into microbial bodies thus again reducing NH_4^+ availability to nitrifying bacteria. Finally, based on the assumption that plants benefit from reduced nitrification, it has long been hypothesized that plants (both litter and roots) chemically inhibit nitrifiers. However, unequivocal data demonstrating lower gross rates of nitrification in rhizosphere soil have been lacking. We measured gross rates of nitrification in rhizosphere soil from *Avena barbata*, a common annual grass in California using both microcosm and field experiments (Herman *et al.* 2006, Hawkes *et al.* 2005). In the microcosm experiment, actual gross rates of nitrification in the rhizosphere were zero in areas of active NH_4^+ uptake by the root (8–16 cm from root tip); in areas from which little NH_4^+ uptake was occurring (0–8 cm from root tip), rates of nitrification were indistinguishable from those of bulk soil. Thus root competition for NH_4^+ can substantially reduce nitrification in zones of active root uptake. In the field experiment, we observed that rates of gross nitrification were higher in the presence of *A. barbata* roots than in the presence of a complex plant community; thus plant community composition can also impact rate of nitrification. Populations of nitrifiers paralleled rates of nitrification in both experiments. In the microcosm study, nitrification potentials were slightly higher in soil adjacent to the 0–8 cm zone compared to those in bulk soil, but lower in soil from the 8–16 cm root zone. In the field experiment, qPCR revealed a direct relationship between the abundance of nitrifiers and gross rates of nitrification.

Denitrification

Denitrification occurs when nitrate is used as an alternative terminal electron acceptor under conditions of oxygen limitation. When root respiration depletes local concentrations of O_2, nitrate reduction in the rhizosphere increases. When root water uptake increases diffusional resupply of O_2 to the rhizosphere compartment, denitrification can be reduced. When root uptake of NO_3 reduces NO_3 availability, denitrification can decrease (Firestone 1982). Thus there is no simple, uniform response of denitrification to the presence of roots. Denitrification may have another important role in rhizosphere processes due to the gaseous intermediates formed during NO_3 reduction to N_2. These intermediates, especially NO, are biologically active and may play a role in seed germination, root growth, and immune response to plant pathogens (Stohr and Ulrich 2002). While NO is known to affect these aspects of plant biology, the extent to which microbially generated NO plays a role in these processes is unclear.

LINKING RHIZOSPHERE MICROBIOLOGY WITH LARGER SCALES

Though we know that many ecosystem processes are under microbial control, simple models of these processes, including nitrogen mineralization, generally work well without explicitly including soil microbes. Nevertheless, microbial response data are crucial to the success of these models – starting conditions, response functions, and parameter values are all developed from biological data (Andrén *et al.* 1999). There are some conditions where inclusion of microbial mechanisms behind ecosystem processes may be critical, particularly under scenarios where a change in the microbial community could feedback to directly change ecosystem process rates.

We have discussed two scenarios in which the composition and interactions of the rhizosphere microbial community may impact ecosystem nutrient cycling. In one scenario, communication among rhizosphere microbes and between roots and microbes can impact gross rates of N-mineralization and hence the availability of a potentially limiting nutrient. In a second scenario, differences in the community composition of microbes mediating a process can directly impact the rates and characteristics of that process. Nitrification and denitrification rates have been shown to differ in proportion to the abundance and composition of nitrifying and denitrifying communities in soil (e.g., Cavigelli and Robertson 2000, Hawkes *et al.* 2005). In contrast, microbial functional redundancy in processes such as mineralization suggests that the specific composition or diversity of the microbial community should not substantially affect the function (Girvan *et al.* 2005). The development of stable isotope probing (Radajewski *et al.* 2002) may allow a greater understanding of the role of specific microbial populations in processes performed by diverse groups of microorganisms.

Global changes in climate and plant communities may further alter microbial communities with consequences for ecosystem process rates. Some predicted global change, however, may not directly affect the composition and function of bacteria in the rhizosphere. For example, rhizosphere microbes experience higher concentrations of CO_2 in soil than are found in the atmosphere, making increased atmospheric CO_2 unlikely to directly affect these communities. Indirectly, however, the resulting increases in plant allocation of carbon belowground could change both microbial abundance and function (Hu *et al.* 2001).

Direct effects on rhizosphere microbial communities are likely to occur when plant community composition is altered, through processes such as exotic plant invasions and land use change. Invasion of novel habitats by exotic plant species can alter microbial communities as well as microbially mediated processes. Already we know that successful invaders tend to have fewer fungal pathogens (Mitchell and Power 2003) and that plant invasions can affect communities of bacteria and mycorrhizal fungi, soil aggregation, gross rates of nitrification, and soil enzymes (Andrade *et al.* 1998b, Kourtev *et al.* 2003, Hawkes *et al.* 2005, 2006). Based on these studies and the results of our phylogenetic analysis, we predict that woody shrub and tree invasions of grass or forb-dominated areas are likely to cause the largest shifts in microbial community composition. Land use changes, including those that remove the woody layer from a plant community, are also likely to dramatically alter microbial community composition – some evidence for this already exists in agricultural systems. More on the predicted effects of global changes on mycorrhizal fungi are discussed in Chapter 4.

1.5   CLOSING OBSERVATIONS

Research since the 1990s has underscored the fact that understanding and quantifying the interactions among plants and soil microbes is essential to understanding both plant and soil microbial community ecology and the roles that these communities play in ecosystem function.

Tremendous progress has recently been made in the area of plant–soil microbial interactions and there now exist powerful new tools that show promise for rapid and continued expansion of our understanding. We know that microbial communities in rhizosphere soils differ from those of surrounding soil and we know that the rates and characteristics of N- and C-cycling processes differ in these soils. But we are still largely unable to link differences in rhizosphere microbial communities to differences in nutrient cycling and ultimately in ecosystem function. We are just beginning to understand how and to what degree plants impact bulk soil communities and rhizosphere communities. What is the relative importance of the plant versus the soil environment in framing the microbial composition of rhizosphere communities? How and to what degree do soil microbial communities impact plant physiological, population, and community ecology? Exactly how do plant–microbial feedback loops work? Over what time frame do these interactions develop? How does the chemical–physical environment of soil impact these biotic interactions? What are the roles of signaling and cell–cell communication in mediating root–microbial interactions? Researchers equipped to address these complex questions need expertise in a breath of areas ranging from pedology to biometeorology to plant physiology to microbial genetics. Thus we expect that the most substantial advances in this area will be made by research collaborations that encompass a range of disciplinary expertise.

ACKNOWLEDGEMENTS

We thank Zoe Cardon, Julie Whitbeck, and two anonymous reviewers for providing comments on the manuscript. Phil Hugenholz and Brett Baker provided help with the phylogenetic analysis. The work was supported by a CA AES Project 6117-H. Hawkes was supported by an NSF Microbial Biology Postdoctoral Fellowship (DBI-0200720) and DeAngelis by an EPA-STAR graduate fellowship.

REFERENCES

Allen, M.F. 1991. *The Ecology of Mycorrhizae*. Cambridge University Press, Cambridge.

Andrade, G., R.G. Linderman, and G.J. Bethlenfalvay. 1998a. Bacterial associations with the mycorrhizosphere and hyphosphere of the arbuscular mycorrhizal fungus *Glomus mosseae*. *Plant and Soil* 202:79–87.

Andrade, G., K.L. Mihara, R.G. Linderman, and G.J. Bethlenfalvay. 1998b. Soil aggregation status and rhizobacteria in the mycorrhizosphere. *Plant and Soil* 202:89–96.

Andrén, Q., L. Brussaard, and M. Clarholm. 1999. Soil organism influence on ecosystem-level processes – bypassing the ecological hierarchy? *Applied Soil Ecology* 11:177–188.

Ashelford, K.E., M.J. Day, and J.C. Fry. 2003. Elevated abundance of bacteriophage infecting bacteria in soil. *Applied and Environmental Microbiology* 69:285–289.

Badalucco, L., P.J. Kuikman, and P. Nannipieri. 1996. Protease and deaminase activities in wheat rhizosphere and their relation to bacterial and protozoan populations. *Biology and Fertility of Soils* 23:99–104.

Barazani, O., and J. Friedman. 2001. Allelopathic bacteria and their impact on higher plants. *Critical Reviews in Microbiology* 27:41–55.

Belnap, J., C.V. Hawkes, and M.K. Firestone. 2003. Boundaries in miniature: two examples from soil. *Bioscience* 53:739–749.

Berg, G., N. Roskot, A. Steidle, L. Eberl, A. Zock, and K. Smalla. 2002. Plant-dependent genotypic and phenotypic diversity of antagonistic rhizobacteria isolated from different *Verticillium* host plants. *Applied and Environmental Microbiology* 68:3328–3338.

Bever, J.D., K.M. Westover, and J. Antonovics. 1997. Incorporating the soil community into plant population dynamics: the utility of the feedback approach. *Journal of Ecology* 85:561–573.

Bevivino, A., S. Sarrocco, C. Dalmastri, S. Tabacchioni, C. Cantale, and L. Chiarini. 1998. Characterization of a free-living maize-rhizosphere population of Burkholderia cepacia: effect of seed treatment on disease suppression and growth promotion of maize. *FEMS Microbiology Ecology* 27:225–237.

Bianciotto, V., E. Lumini, L. Lanfranco, D. Minerdi, P. Bonfante, and S. Perotto. 2000. Detection and identification of bacterial endosymbionts in arbuscular mycorrhizal fungi belonging to the family Gigasporaceae. *Applied and Environmental Microbiology* 66:4503–4509.

Caldwell, M.M., and J.H. Richards. 1989. Hydraulic lift – water efflux from upper roots improves effectiveness of water-uptake by deep roots. *Oecologia* 79:1–5.

Campbell, C.D., S.J. Grayston, and D.J. Hirst. 1997. Use of rhizosphere carbon sources in sole carbon source tests to discriminate soil microbial communities. *Journal of Microbiological Methods* 30:33–41.

Carelli, M., S. Gnocchi, S. Fancelli, A. Mengoni, D. Paffetti, C. Scotti, and M. Bazzicalupo. 2000. Genetic diversity and dynamics of Sinorhizobium meliloti populations nodulating different alfalfa cultivars in Italian soils. *Applied and Environmental Microbiology* 66:4785–4789.

Cavigelli, M.A., and G.P. Robertson. 2000. The functional significance of denitrifier community composition in a terrestrial ecosystem. *Ecology* 81:1402–1414.

Chelius, M.K., and E.W. Triplett. 2001. The diversity of archaea and bacteria in association with the roots of *Zea mays* L. *Microbial Ecology* 41:252–263.

Chow, M.L., C.C. Radomski, J.M. McDermott, J. Davies, and P.E. Axelrood. 2002. Molecular characterization of bacterial diversity in Lodgepole pine (*Pinus contorta*) rhizosphere soils from British Columbia forest soils differing in disturbance and geographic source. *FEMS Microbiology Ecology* 42:347–357.

Crespi, B.J. 2001. The evolution of social behavior in microorganisms – Response from Crespi. *Trends in Ecology and Evolution* 16:607–607.

Dalmastri Cfiore, A., C. Alisi, A. Bevivino, S. Tabacchioni, G. Giuliano, A.R. Sprocati, L. Segre, E. Mahenthiralingam, L. Chiarini, and P. Vandamme. 2003. A rhizospheric Burkholderia cepacia complex population: genotypic and phenotypic diversity of *Burkholderia cenocepacia* and *Burkholderia ambifaria*. *FEMS Microbiology Ecology* 46:179–187.

Duineveld, B.M., G.A. Kowalchuk, A. Keijzer, J.D. van Elsas, and J.A. van Veen. 2001. Analysis of bacterial communities in the rhizosphere of chrysanthemum via denaturing gradient gel electrophoresis of PCR-amplified 16S rRNA as well as DNA fragments coding for 16S rRNA. *Applied and Environmental Microbiology* 67:172–178.

Firestone, M.K. 1982. Biological denitrification. In: Stevenson, F.J. (ed.) Nitrogen in Agricultural Soils. *American Society of Agronomy*, Madison, WI, pp. 289–326.

Garbaye, J. 1991. Biological interactions in the mycorrhizosphere. *Experientia* 47:370–375.

Girvan, M.S., C.D. Campbell, K. Killham, J.I. Prosser, and L.A. Glover. 2005. Bacterial diversity promotes community stability and functional resilience after perturbation. *Environmental Microbiology* 7:301–313.

González-Andrés, F., and J-M. Ortiz. 1999. Specificity of rhizobia nodulating *Genista monspessulana* and *Genista linifolia* in vitro and in field situations. *Arid Soil Research Rehabilitation* 13:223–237.

Halverson, L.J., T.M. Jones, and M.K. Firestone. 2000. Release of intracellular solutes by four soil bacteria exposed to dilution stress. *Soil Science Society of America Journal* 64:1630–1637.

Hawkes, C.V., I. Wren, D. Herman, and M.K. Firestone. 2005. Plant invasion alters nitrogen cycling by modifying soil microbial communities. *Ecology Letters* 8:976–985.

Hedges, R.W., and E. Messens. 1990. Genetic aspects of rhizosphere interactions. In: Lynch, J.M. (ed.) The Rhizosphere. John Wiley & Sons, New York, pp. 59–97.

Herman, D.J., K.K. Johnson, C.H. Jaeger, E. Schwartz, and M.K. Firestone. 2006. Root influence on nitrogen mineralization and nitrification in rhizosphere soil of slender wild oats. *Soil Science Society of America Journal* 70:1504–1511.

Hinsinger, P., C. Plassard, C.X. Tang, and B. Jaillard. 2003. Origins of root-mediated pH changes in the rhizosphere and their responses to environmental constraints: a review. *Plant and Soil* 248:43–59.

Hu, S., F.S. Chapin, M.K. Firestone, C.B. Field, and N.R. Chiariello. 2001. Nitrogen limitation of microbial decomposition in a grassland under elevated CO_2. *Nature* 409:188.

Jaeger, C.H., S.E. Lindow, S. Miller, E. Clark, and M.K. Firestone. 1999. Mapping of sugar and amino acid availability in soil around roots with bacterial sensors of sucrose and Tryptophan. *Applied and Environmental Microbiology* 65:2685–2690.

Kaiser, O., A. Puhler, and W. Selbitschka. 2001. Phylogenetic analysis of microbial diversity in the rhizoplane of oilseed rape (Brassica napus cv. Westar) employing cultivation-dependent and cultivation-independent approaches. *Microbial Ecology* 42:136–149.

Kaye, J.P., and S.C. Hart. 1997. Competition for nitrogen between plants and soil microorganisms. *Trends in Ecology and Evolution* 12:139–143.

Khan, M.R., and S.M. Khan. 2002. Effects of root-dip treatment with certain phosphate solubilizing microorganisms on the fusarial wilt of tomato. *Bioresource Technology* 85:213–215.

Kieft, T.L., E. Soroker, and M.K. Firestone. 1987. Microbial biomass response to a rapid increase in water potential when dry soil is wetted. *Soil Biology and Biochemistry* 19:119–126.

Klemedtsson, L., P. Berg, M. Clarholm, J. Schnurer, and T. Rosswall. 1987. Microbial nitrogen transformations in the root environment of barley. *Soil Biology and Biochemistry* 19:551–558.

Kourtev, P.S., J.G. Ehrenfeld, and M. Haggblom. 2003. Experimental analysis of the effect of exotic and native plant species on the structure and function of soil microbial communities. *Soil Biology and Biochemistry* 35:895–905.

Kuske, C.R. et al. 2002. Comparison of soil bacterial communities in rhizospheres of three plant species and the interspaces in an arid grassland. *Applied and Environmental Microbiology* 68:1854–1863.

Loh, J., E.A. Pierson, L.S. Pierson, G. Stacey, and A. Chatterjee. 2002. Quorum sensing in plant-associated bacteria. *Current Opinion in Plant Biology* 5:285–290.

Ludwig, W., O. Strunk, R. Westram, H. Meier, Yadhukumar, A. Buchner, T. Lai, S. Steppi, G. Jobb, W. Förster, I. Brettske, S. Gerber, A.W. Ginhart, O. Gross, S. Grumann, S. Hermann, R. Jost, A. König, T. Liss, R. Lüßmann, M. May, B. Nonhoff, B. Reichel, R. Strehlow, A. Stamatakis, N. Stuckmann, A. Vilbig, M. Lenke, T. Ludwig, A. Bode, and K-H. Schleifer. 2004. ARB: a software environment for sequence data. *Nucleic Acids Research* 32:1363–1371.

Madigan, M.T., J.M. Martinko, and J. Parker. 2003. *Brock Biology of Microorganisms*, 10th Edition. Prentice Hall, Upper Saddle River, NJ.

Marilley, L., and M. Aragno. 1999. Phylogenetic diversity of bacterial communities differing in degree of proximity of *Lolium perenne* and *Trifolium repens* roots. *Applied Soil Ecology* 13:127–136.

Marschner, P., D.E. Crowley, and R.M. Higashi. 1997. Root exudation and physiological status of a root-colonizing fluorescent pseudomonad in mycorrhizal and non-mycorrhizal pepper (Capsicum annum L.). *Plant and Soil* 189:11–20.

Miethling, R., K. Ahrends, and C.C Tebbe. 2003. Structural differences in the rhizosphere communities of legumes are not equally reflected in community-level physiological profiles. *Soil Biology and Biochemistry* 35:1405–1410.

Miller, M.B., and B.L. Bassler. 2001. Quorum sensing in bacteria. *Annual Review of Microbiology* 55:165–199.

Misko, A.L., and J.J. Germida. 2002. Taxonomic and functional diversity of pseudomonads isolated from the roots of field-grown canola. *FEMS Microbiology Ecology* 42:399–407.

Mitchell, C.E., and A.G. Power. 2003. Release of invasive plants from fungal and viral pathogens. *Nature* 421:625–627.

Murray, M.S., and E.M. Hansen. 1997. Susceptibility of Pacific yew to *Phytophthora lateralis*. *Plant Disease* 81:1400–1404.

Nehl, D.B., S.J. Allen, and J.F. Brown. 1997. Deleterious rhizosphere bacteria: an integrating perspective. *Applied Soil Ecology* 5:1–20.

Neumann, G., and V. Romheld. 2001. The release of root exudates as affected by the plant's physiological status. In: Pinton, R., Varanini, Z., and Nannipieri, P. (eds) The rhizosphere: biochemistry and organic substances at the soil-plant interface. Marcel-Dekker, Inc., New York, pp. 41–93.

Normander, B., and J.I. Prosser. 2000. Bacterial origin and community composition in the barley phytosphere as a function of habitat and presowing conditions. *Applied and Environmental Microbiology* 66:4372–4377.

Olff, H., B. Hoorens, R.G.M. de Goede, W.H. van der Putten, and J.M. Gleichman. 2000. Small-scale shifting mosaics of two dominant grassland species: The possible role of soil-borne pathogens. *Oecologia* 125:45–54

Packer, A., and K. Clay. 2003. Soil pathogens and *Prunus serotina* seedling and sapling growth near conspecific trees. *Ecology* 84:108–119.

Papendick, R.I., and G.S. Campbell. 1975. Water potential in the rhizosphere and plant and methods of measurement and experimental control. In: Bruehl, G.W. (ed.) Biology and control of soil-born plant pathogens. American Phytopathological Society, St. Paul, MN, pp. 34–49.

Priha, O., T. Hallantie, and A. Smolander. 1999. Comparing microbial biomass, denitrification enzyme activity, and numbers of nitrifiers in the rhizospheres of *Pinus sylvestris, Psicea abies*, and *Betula pendula* seedlings by microscale methods. *Biology and Fertility of Soils* 30:14–19.

Prosser, J.I. 2002. Molecular and functional diversity in soil micro-organisms. *Plant and Soil* 244:9–17.

Radajewski, S., G. Webster, D.S. Reay, S.A. Morris, P. Ineson, D.B. Nedwell, J.I. Prosser, and J.C. Murrell. 2002. Identification of active methylotroph populations in an acidic forest soil by stableisotope probing. *Microbiology-SGM* 148:2331–2342.

Rainey, P.B. 1999. Adaptation of Pseudomonas fluorescens to the plant rhizosphere. *Environmental Microbiology* 1:243–257.

Reynolds, H.L., A. Packer, J.D. Bever, and K. Clay. 2003. Grassroots ecology: plant-microbe-soil interactions as drivers of plant community structure and dynamics. *Ecology* 84:2281–2291.

Schimel, J.P., and J. Bennett. 2004. Nitrogen mineralization: challenges of a changing paradigm. *Ecology* 85:591–602.

Schmalenberger, A., and C.C. Tebbe. 2002. Bacterial community composition in the rhizosphere of a transgenic, herbicide-resistant maize (*Zea mays*) and comparison to its non-transgenic cultivar Biosphore. *FEMS Microbiology Ecology* 40:29–37.

Schmalenberger, A., and C.C. Tebbe. 2003a. Genetic profiling of noncultivated bacteria from the rhizospheres of sugar beet (*Beta vulgaris*) reveal field and annual variability but no effect of a transgenic herbicide resistance. *Canadian Journal of Microbiology* 49:1–8.

Schmalenberger, A., and C.C. Tebbe. 2003b. Bacterial diversity in maize rhizospheres: conclusions on the use of genetic profiles based on PCR-amplified partial small subunit rRNA genes in ecological studies. *Molecular Ecology* 12:251–261.

Schmidt, W., W. Michalke, and A. Schikora. 2003. Proton pumping by tomato roots. Effect of Fe deficiency and hormones on the activity and distribution of plasma membrane H^+-ATPase in rhizodermal cells. *Plant Cell and Environment* 26:361–370.

Schwab, S.M., R.T. Leonard, and J.A. Menge. 1984. Quantitative and qualitative comparison of root exudates of mycorrhizal and nonmycorrhizal plant species. *Canadian Journal of Botany-Revue Canadienne De Botanique* 62:1227–1231.

Smalla, K., G. Wieland, A. Buchner, A. Zock, J. Parzy, S. Kaiser, N. Roskot, H. Heuer, and G. Berg. 2001. Bulk and rhizosphere soil bacterial communities studied by denaturing gradient gel electrophoresis: plant-dependent enrichment and seasonal shifts revealed. *Applied and Environmental Microbiology* 67:4742–4751.

Smith, K.P., J. Handelsman, and R.M. Goodman. 1999. Genetic basis in plants for interactions with disease- suppressive bacteria. *Proceedings of the National Academy of Sciences of the United States of America* 96:4786–4790.

Sorensen, J. 1997. The rhizosphere as a habitat for soil microorganisms. In: van Elsas, J.D., Trevors, J.T., and Wellington, E.M.H. (eds) *Modern Soil Microbiology*. Marcel-Dekker, Inc., New York, pp. 21–45.

Stephan, A., A.H. Meyer, and B. Schmid. 2000. Plant diversity affects culturable soil bacteria in experimental grassland communities. *Journal of Ecology* 88:988–998.

Stohr, C., and W.R. Ullrich. 2002. Generation and possible roles of NO in plant roots and their apoplastic space. *Journal of Experimental Botany* 53:2293–2303.

Teplitski, M., J.B. Robinson, and W.D. Bauer. 2000. Plants secrete substances that mimic bacterial N-acyl homoserine lactone signal activities and affect population density-dependent behaviors in associated bacteria. *Molecular Plant Microbe Interations* 13:637–648.

Tesfaye, M., N.S. Dufault, M.R. Dornbusch, D.L. Allan, C.P. Vance, and D. Samac. 2003. Influence of enhanced malate dehydrogenase expression by alfalfa on diversity of rhizobacteria and soil nutrient availability. *Soil Biology and Biochemistry* 35:1103–1113.

Thrall, P.H., J.J. Burdon, and M.J. Woods. 2000. Variation in the effectiveness of symbiotic associations between native rhizobia and temperate Australian legumes: interactions within and between genera. *Journal of Applied Ecology* 37:52–65.

Weste, G. 1986. Vegetation changes associated with invasion by Phytophthora-cinnamomi of defined plots in the Brisbane Ranges, Victoria, 1975–1985. *Australian Journal of Botany* 34:633–648.

Weste, G., and D.H. Ashton. 1994. Regeneration and survival of indigenous dry sclerophyll species in the Brisbane Ranges, Victoria, after a *Phytophthora cinnamomi* epidemic. *Australian Journal of Botany* 42:239–253.

Wilson, G.W.T., D.C. Hartnett, M.D. Smith, and K. Kobbeman. 2001. Effects of mycorrhizae on growth and demography of tallgrass prairie forbs. *American Journal of Botany* 88:1452–1457.

Yang, C.H., and D.E. Crowley. 2000. Rhizosphere microbial community structure in relation to root location and plant iron nutritional status. *Applied and Environmental Microbiology* 66: 345–351.

Yang, C.H., D.E. Crowley, and J.A. Menge. 2001. 16S rDNA fingerprinting of rhizosphere bacterial communities associated with healthy and *Phytophthora* infected avocado roots. *FEMS Microbiology Ecology* 35:129–136.

CHAPTER 2

Carbon Fluxes in the Rhizosphere

Weixin Cheng and Alexander Gershenson

2.1 INTRODUCTION

Terrestrial ecosystems are intimately connected to atmospheric CO_2 levels through photosynthetic fixation of CO_2, sequestration of C into biomass and soils, and the subsequent release of CO_2 through respiration and decomposition of organic matter. Considering all the pools and fluxes of C within ecosystems, C-cycling belowground is increasingly being recognized as one of the most significant components of the carbon cycle (e.g., Zak and Pregitzer 1998). Globally, the input of C to the soil has been estimated to be as great as 60×10^{15} g yr^{-1}, approximately one order of magnitude larger than the global annual rate of fossil fuel burning and other anthropogenic emissions, which is at 6×10^{15} g yr^{-1} currently (Post *et al.* 1990). Thus, small changes in the equilibrium between sinputs and decomposition could have a significant impact on atmospheric CO_2 concentrations, which may either exacerbate or reduce the consequence of burning of fossil fuels (Schimel 1995). Belowground CO_2 efflux can be partitioned into two distinct processes: (1) rhizosphere respiration or root-derived CO_2, including root respiration and microbial respiration utilizing materials released from live roots and (2) microbial decomposition of soil organic matter (SOM), or soil-derived CO_2. While the two processes act separately, they may also be linked through rhizosphere interactions, which may exert a stimulative (priming effect) or a suppressive influence on SOM decomposition (Cheng 1999). As a measure of main energy use for the acquisition of belowground resources (e.g., nutrients and water), rhizosphere respiration may range from 30 to 80 percent of total belowground CO_2 efflux (Hanson *et al.* 2000) in various ecosystems. Root-associated C fluxes represent a major portion of the input to and the output from the belowground C pool (Schimel 1995).

The rhizosphere harbors very high numbers and activities of organisms. Concentrations of microbes in the rhizosphere can reach 10^{10}–10^{12} per gram of rhizosphere soil as compared to often $<10^8$ in the bulk soil (Foster 1988). Invertebrate density in the rhizosphere is at least two orders of magnitude greater than in the bulk soil. This highly active system associated with plant roots is mainly supported by the carbon input from live roots, which may include sloughed-off materials (cells and mucilage), dead root hairs, and root exudates. This input, along with root turnover, may account for up to 50 percent of the net primary production in various ecosystems (Whipps 1990). The flow of energy and the function of this carbon flux within ecosystems constitute a major area of interest in ecology. To understand the interactions between the three biotic components of the soil, that is roots, microflora and fauna, a necessary first step is to determine how much organic material is contributed to the soil by roots.

In order to facilitate discussion on the various kinds of carbon input into the rhizosphere from roots, we need to first briefly describe the main categories of rhizosphere carbon input and their relationships (Figure 2.1). Among the plant-derived carbon allocated belowground via roots, there are three main components:

1. Roots mass, either alive or dead, which can be normally assessed by physical sampling.
2. Other materials of plant origin remaining in the rhizosphere or the surrounding soil, often called rhizodeposits, which are readily utilized

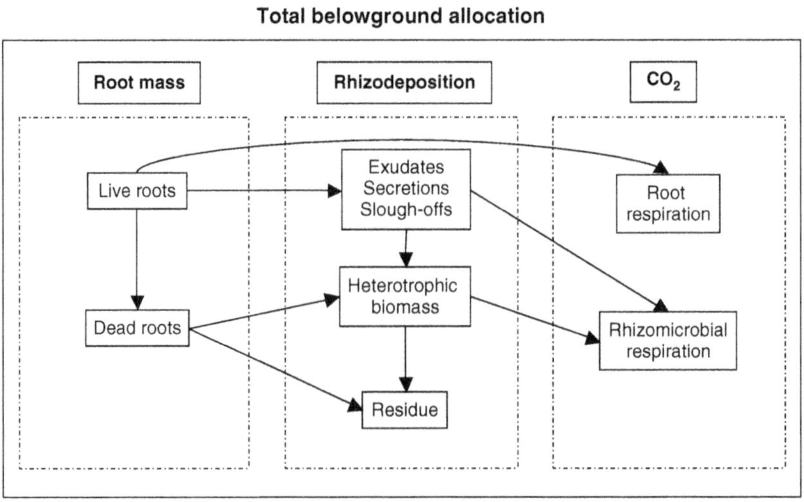

FIGURE 2.1 Main categories of rhizodeposits and their interrelationships.

and transformed by rhizosphere biota and simultaneously mixed with soil organic materials.

3. Carbon dioxide either from respiration of roots and root symbionts, such as mycorrhizae and nodules, or from rhizosphere microbial respiration utilizing root-derived substrates.

Therefore, studying carbon fluxes in the rhizosphere requires investigation of all three categories mentioned above, in addition to aboveground components. Input by root growth and turnover is extensively covered in Chapter 7. This chapter will primarily focus on the second and the third categories – rhizodeposition and CO_2 efflux. Because these materials are intimately mixed with soil-derived carbon sources and simultaneously transformed by soil microorganisms, investigating these carbon fluxes requires a suite of methods that either eliminate the soil components, such as using a sterile liquid culture technique, or are capable of tracing root-derived sources separately from soil-derived sources, such as isotope labeling. One of the greatest challenges in rhizosphere research, dictated in large part by the nature of the medium itself, is observation of processes in situ. The methodologies currently available to us often do not allow for such direct observation; however, existing and recently developed methods offer us the opportunity to examine rhizosphere processes with ever-increasing sophistication and approximation of actual soil conditions.

2.2   QUANTITY AND QUALITY OF RHIZODEPOSITS

Rhizodeposition was first defined by Whipps and Lynch (1985) as all material lost from plant roots, including water-soluble exudates, secretions of insoluble materials, lysates, dead fine roots, and gases, such as CO_2 and ethylene. Because several reviews on this topic have been published (e.g., Whipps 1990; Kuzyakov and Domanski 2000, Nguyen 2003), only a brief summary is included in this chapter.

The sources of organic C-input from roots can be divided into two main groups: (1) water-soluble exudates, for example sugars, amino acids, organic acids, and so on; and (2) water-insoluble materials, for example sloughed cells and mucilage. Materials in the first group are rapidly metabolized by rhizosphere microorganisms. There are three sources of CO_2 released by a system of living roots and soil: (1) root respiration; (2) microbial respiration utilizing root-derived materials (rhizo-microbial respiration); and (3) microbial respiration using original soil carbon. This intimate association of root respiration and exudation with rhizo-microbial respiration has made studies of root respiration, root exudation, and rhizo-microbial respiration in natural soils very difficult.

The quality or chemical composition of rhizodeposits is an important determinant of the functions and ecological consequences of rhizodeposition. Our understanding of the chemical composition of root exudates and other rhizodeposits is virtually all based on data from experiments using sterile liquid culture methods. Much of the literature on this topic has been reviewed previously (e.g., Whipps 1990). Virtually all kinds of plant molecules and materials can be found in rhizodeposits, although lower molecular weight compounds seem to dominate. Even though labeling methods have been increasingly used to study rhizospheric carbon fluxes, our understanding of the chemical composition of rhizodeposits has not advanced much in the past few decades due to the limitation of available methods. Although the use of Gas Chromatography–Mass Spectrometry (GC–MS) analysis for identification of compounds exuded by roots (Bertin *et al.* 2003) offers potentially exciting developments in this area, such methodologies still depend heavily on significant simplifications of the rhizosphere system. Studying the chemical composition of rhizodeposits under real soil conditions remains a challenge because of the intimate coupling of microbial utilization and transformation of rhizodeposits.

To provide some examples for the strong effect of the particular methodologies used, if we consider that in the case of using nutrient solution cultures under gnotobiotic conditions, the amount of rhizodeposition has been quantified to be less than $0.6 \, mg \, g^{-1}$ of root dry weight (Lambers 1987) for seedlings a few weeks old. Due to these highly artificial conditions, this value must represent considerable underestimation. By increasing complexity within the experimental system, for instance by using solid media and adding microorganisms, the amount of rhizodeposition significantly increases (e.g., Barber and Gunn 1974). Using ^{14}C-labeling techniques, rhizodeposition has been quantified under more realistic conditions. Values of rhizodeposition, measured by this labeling technique, may range between 30 and 90 percent of the carbon transferred to belowground components of various plant–soil systems (Whipps 1990). Differences in soil type, plant species, plant growth stage, and other experimental conditions employed in various studies may have caused the wide range. For young plants of wheat, barley, or pasture grasses, approximately 20–50 percent of net assimilated carbon is transferred into belowground components, including root biomass ($\sim$50%), rhizosphere-derived CO_2 ($\sim$30%), and soil residues ($\sim$20%) (Kuzyakov and Domanski 2000). Given that these percentages are averaged across results from labeling experiments using mostly young plants and various experimental conditions, the distribution pattern for carbon allocated belowground seems relatively consistent. However, the relative distribution among belowground components is most likely time-dependent, because of carbon transfer from root biomass to CO_2 and soil residues via root turnover as the plant ages. The results mentioned above are mostly drawn from experiments using young plants under laboratory conditions. Further studies with longer time span and under settings that bring us closer to in situ conditions are needed to address this issue.

Realistic assessment of the quantity of rhizodeposits in ecosystems remains a challenging task. In the review by Nguyen (2003), data from experiments using both continuous labeling and pulse labeling techniques are compiled and analyzed. The meta-analysis suggests that the total quantity of rhizodeposits is significantly influenced by plant species, plant age, the presence or absence of rhizosphere microorganisms, soil texture, and nitrogen availability. Based on a recent study of 12 Mediterranean species of herbaceous plants (Warembourg *et al.* 2003), the percentage of assimilated C allocated to belowground differs significantly between major groups of species (i.e., grasses, legumes, and non-legume forbs), but not significantly different between plant species within each group. Less carbon is allocated belowground as plants age, based on data mostly from annual plant species. The presence of rhizosphere microorganisms substantially increases the quantity of rhizodeposits, as compared to sterile cultures, indicating that sterile cultures should not be used to realistically quantify total rhizodeposits or total carbon allocation to belowground components, although such techniques may allow a qualitative assessment of the types of compounds exuded. Relatively higher (up to 15%) soil clay contents seem to increase total rhizodeposits; however, nitrogen fertilization is likely to significantly reduce total rhizodeposits. As clearly pointed out in the review by Farrar and Jones (2000), plant carbon allocation imposes the first level of control on the total quantity of rhizodeposits. Therefore, any biotic or environmental conditions that may affect plant carbon allocation will exert controls on rhizodeposition. For example, pulse growth of aboveground components of *Quercus rubra* seems to vary inversely with rhizosphere respiration, supposedly due to the change of plant C allocation between aboveground and belowground components (Cardon *et al.* 2002).

A cautionary note is necessary for a reliable use of the above-mentioned results for the assessment of the quantity of rhizodeposits. The majority of the data is taken from experiments of relatively short duration with young plants. Therefore, the interpretation of the data should be limited to such circumstances. The quantitative assessment of rhizodeposits is also method-dependent (Whipps 1990; Kuzyakov and Domanski 2000). Method limitations are considered later in a separate section of this chapter.

2.3 RHIZOSPHERE CARBON FLUXES UNDER ELEVATED CO_2

Plants grown under elevated CO_2 conditions often exhibit increased growth and a disproportional increase in C allocation to roots (Norby et al. 1986; Matamala and Schlesinger 2000; Norby et al. 2002), total rhizosphere respiration (Cheng et al. 2000), and rhizodeposition (Kuikman et al. 1991; Billes et al. 1993). By using carbon isotope tracers in CO_2 enrichment experiments at the small-pot scale, several studies have demonstrated that, compared to ambient CO_2 levels, elevated CO_2 increased the amount of carbon allocated to the rhizosphere by enhanced root deposition or total rhizosphere respiration (Hungate et al. 1997; Cheng and Johnson 1998). In general, total carbon input to the rhizosphere is significantly increased when plants are grown under elevated CO_2 (Table 2.1) (also see Chapter 7).

The degree of CO_2 enhancement of rhizosphere respiration could be much higher than enhancement of root biomass (Table 2.2). In a continuous ^{14}C-labeling study using wheat, Lekkerkerk et al. (1990) reported that the wheat plants grown under the elevated CO_2 treatment produced 74 percent more rhizosphere-respired C and only 17 percent more root biomass compared to the ambient treatment. Hungate et al. (1997), in a microcosm experiment with mixed grasses, reported that elevated CO_2 enhanced total rhizosphere deposition by 56 percent and root biomass by less than 15 percent. Cheng and Johnson (1998) reported that, compared with the ambient CO_2 treatment, wheat rhizosphere respiration rate increased 60 percent and root biomass increased only 26 percent under the elevated CO_2 treatment. Two potential mechanisms could be posited as potential causes of these results. First, roots grown under elevated CO_2 exuded more and had higher turnover rates than roots grown under the ambient treatment, resulting in a more than proportional increase in total rhizosphere respiration under elevated CO_2. Second, rhizosphere microbial associations were more enhanced under elevated CO_2 than under ambient CO_2, resulting in higher rhizosphere microbial activities per unit of root growth.

Some evidence supports the first hypothesis. Using the isotopic trapping method (Cheng et al. 1993, 1994), approximately a 60 percent increase in soluble C concentration was found in the rhizosphere when wheat plants were grown under elevated CO_2 compared to ambient CO_2, indicating that roots grown under elevated CO_2 exuded more soluble C (Cheng and Johnson 1998). Pregitzer et al.'s (see Chapter 7) review evidences that root turnover rates are higher for plants grown under elevated CO_2 than under ambient. However, the amount of extra C input to the rhizosphere due to the enhanced root turnover under elevated CO_2 was expected to be low in most tracer studies of short duration, since the life span of the roots was probably longer than

TABLE 2.1 Effect of Elevated CO_2 on Rhizodeposition

Plant species	Conditions	Rhizo-deposition	Rhizo CO_2	Soil residue	SRD*	Reference
Wheat	Microcosm, ^{14}C continuous	↑	↑	↑	↑	Kuikman et al. 1991
Wheat	Microcosm, ^{14}C continuous	↑	↑	↑	NE	Billes et al. 1993
Chestnut	Microcosm, ^{14}C-Pulse	↑	↑	ND	ND	Rouhier et al. 1996
Wheat	Microcosm, ^{14}C-Pulse	↑	↑	↑	ND	Paterson et al. 1996
Rye grass	Microcosm, ^{14}C-Pulse	NE	NE	ND	ND	Paterson et al. 1996
Rye grass	Microcosm, ^{14}C-Pulse	↑	↑	ND	ND	Paterson et al. 1999
Wheat	Microcosm, ^{13}C natural	↑	↑	↑	↑	Cheng and Johnson 1998
Sunflower	Microcosm, ^{13}C natural	↑	↑	NE	↑	Cheng et al. 2000

(NE = no significant effect; ND = not determined; ↑ = significantly increased).
* SRD = specific rhizodeposition, or total deposition g^{-1} of roots.

TABLE 2.2   Percent Increase of Root Biomass and Rhizosphere CO_2
Efflux in Response to Elevated Atmospheric CO_2 Concentrations,
Calculated as (Elevated-Ambient)/Ambient × 100

Root biomass	Rhizosphere CO_2	Reference
17	74	Lekkerkerk *et al.* 1990
15	56	Hungate *et al.* 1997
26	60	Cheng and Johnson 1998
50	96	Cheng *et al.* 2000

the duration of the experiment (Eissenstat and Yanai 1997). Enhanced root exudation was probably the major component of this extra C input to the rhizosphere in these short experiments. Enhanced root turnover under elevated CO_2 for forest ecosystems might contribute more since root turnover was one of the important processes responsible for C input in forests (see Chapter 7). In a deconvolution analysis of soil CO_2 data from the Duke Free-Air CO_2 Enrichment (FACE) experiment, Luo *et al.* (2001) indicated that fine root turnover is a major process adding C to the rhizosphere in response to elevated CO_2, and that root respiration and exudation are less affected by elevated CO_2. The second hypothesis, suggesting that enhancement of rhizosphere respiration under elevated CO_2 is linked to enhanced root–microbial associations, is also supported by evidence in the literature. Elevated CO_2 increased both the percentage of infection of vesicular-arbuscular mycorrhizae and percentage of infection of ectomycorrhizae (see Chapter 4). Elevated CO_2 also increased symbiotic N_2-fixation across several types of associations (Arnone and Gordon 1990; Thomas *et al.* 1991; Tissue *et al.* 1997). However, direct evidence of higher rhizosphere symbiotic activities per unit of root growth under elevated CO_2 is still lacking.

2.4   FUNCTIONAL CONSIDERATIONS

The large quantities of rhizodeposits apparently represent a significant portion of the plant carbon balance and an important source of substrates for soil organisms. In addition to the quantitative significance, their functional significance warrants some attention. What does rhizodeposition do to the plants, to the soil biota, and to the nutrient cycling processes in the soil?

As to the plants, is rhizodeposition a simple passive wasting process, a passive loss of soluble materials by diffusion, an overflow of assimilates when other sinks for photosynthate are limited, active secretion and excretion, or all of the above? Based on data from a liquid culture experiment, root exudation of amino acids and sugars seems to occur passively through diffusion process, and is affected by membrane integrity (Jones and Darrah 1995). Some evidence also suggests that root exudation and respiration may act as overflow mechanisms for excessive photosynthate accumulation (Herald 1980; Lambers 1987). However, many studies in plant nutrition have demonstrated that roots secrete organic acids and other materials for the purpose of nutrient acquisition, such as phosphorus mobilization and iron activation (Marschner 1995).

For soil microflora and fauna, do rhizodeposits act as an important base for the soil food web, or a component of the molecular control points for the coevolution of plants and rhizosphere organisms, or both? Some recent work seems to suggest that rhizodeposits provide the base for a very important part of the soil food web (Garrett et al. 2001; see Chapters 3 and 5). Some of the compounds in rhizodeposits may also act as messengers in regulating the interactions between roots and soil microflora and between different kinds of rhizosphere organisms (Hirsch et al. 2003; Phillips et al. 2003; see Chapters 1 and 3).

Our understanding of the ecological functions of rhizodeposition relies heavily on our ability to study the chemical composition of rhizodeposits. Because the majority of data related to functional understanding are generated from liquid culturing experiments, the applicability of these results to a real soil environment is also limited. The use of reporter genes (Jaeger et al. 1999; Killham and Yeomans 2001; see Chapter 1) may offer new hopes for improvements. For example, the quantity and the chemical forms of some root exudates can be investigated in real soil environment with the help of reporter genes (Jaeger et al. 1999). Evidence from some studies supports a general belief that rhizodeposition exerts strong positive or negative controls on soil organic matter decomposition (Cheng and Kuzyakov 2005). However, very little is known about the role of the rhizosphere effect on decomposition in shaping plant adaptation to various soil environments in the long term. If the rhizosphere effect is closely connected to plant photosynthesis and rhizodeposition (Högberg et al. 2001; Kuzyakov and Cheng 2001), it is conceivable that the rhizosphere effect should be beneficial to plants, and thereby enhance their fitness. Among possible benefits, enhanced nutrient acquisition is often suggested (e.g., Hamilton and Frank 2001). Other benefits may include suppression of root pathogens by supporting healthy microbial communities (e.g., Whipps 2001), conditioning of soil paths for root growth, and improving soil structures and chemical environment, such as pH adjustment (Marschner 1995). If all these benefits occur, the rhizosphere effect on soil organic matter decomposition should be a result of evolutionary processes operating between plants and soil organisms in the overall rhizosphere continuum from incidental to highly symbiotic (see Chapters 1, 3 for more information). Different rhizosphere mechanisms should be selected under different plant and soil environments. This argument seems to be supported

by the fact that different plant–soil couplings produce different rhizosphere
effects on soil organic matter decomposition (Fu and Cheng 2002; Cheng
et al. 2003). Future research is needed to fully illuminate the evolutionary
aspects of rhizodeposition.

2.5 MICROBIAL ASSIMILATION EFFICIENCY OF RHIZODEPOSITS

Microbial carbon assimilation efficiency is commonly defined as microbial
biomass produced as a proportion of total carbon utilized. It is also called
the yield factor. Accumulated evidence suggests that a big proportion of root
exudates is utilized and released as CO_2 in a very short period of time; only
a small portion becomes microbial biomass (Dyer *et al.* 1991; Harris and
Paul 1991). The microbial assimilation efficiency of these exudates (6.5–15%;
Helal and Sauerbeck 1989; Liljeroth *et al.* 1990; Martin and Merckx 1992),
is considerably lower than the theoretical maximum of 60 percent (Payne
1970) and of other sources of carbon in the soil. The microbial assimilation
efficiency is 61 percent for glucose added to the soil after about 40 hours
of incubation (Elliott *et al.* 1983), 27 percent after 61 weeks of incubation
(Johansson 1992), and 47 percent for rye shoots added to the soil after 7 weeks
of incubation (Cheng and Coleman 1990). Why is the microbial assimilation
efficiency of root exudates so low? What mechanisms are there behind this
lower efficiency?

The occurrence of biological N_2-fixation in the rhizosphere may contribute
to the low microbial assimilation efficiency of root exudates. Biological dinitro-
gen fixation requires high amounts of energy, especially those of an associative
nature. At least 16 ATP molecules are consumed to convert one N_2 to two NH_3
molecules, in addition to other processes required for associative N_2-fixation.
If a large proportion of root exudates is used by diazotrophs in the rhizosphere,
the assimilation efficiency of root exudates will be much lower than if it is
being used by non-nitrogen-fixing microbes. Several studies (Liljeroth *et al.*
1990; Van Veen *et al.* 1991) have shown that the assimilation efficiency of
root-derived materials is higher when more nitrogen fertilizer has been used.
It is likely that nitrogen fertilization suppresses diazotrophic activity in the
rhizosphere, which contributes to higher assimilation efficiency. It is widely
known that rhizosphere is one of the important sites for potential associative
free-living nitrogen fixation, due to the favorable conditions in the rhizosphere
(supply of carbon source, mainly root exudates, and the relatively low oxygen
potential caused by root and microbial respiration in the rhizosphere). The list
of free-living nitrogen-fixing bacteria continues to grow as more genera and
species are described. It seems that most plant species in natural environment

are colonized to some degree by free-living diazotrophs (e.g., Kapulnik 1991). The contribution of biologically fixed N_2 by free-living diazotrophs can be substantial in some ecosystems, such as savanna grasslands (Abbadie *et al.* 1992). However, this subject remains controversial in broader perspectives. Some reported values of fixed N_2 by free-living diazotrophs exceed that which can possibly be supported by the estimated amount of carbon available to the diazotrophs. Much of the controversy stems from the energetic requirement of nitrogen fixation process and the estimated amount of carbon available to the rhizosphere diazotrophs (Jones *et al.* 2003). Central to this controversy is the flow of carbon to root-associated diazotrophs in soil-grown plants, since most of the estimates of root exudation are based on gnotobiotic experiments. A better understanding of the contribution of free-living nitrogen-fixing bacteria to the N economy of the rhizosphere requires more suitable methods that allow ecological studies under natural environmental conditions (Kapulnik 1991; Jones *et al.* 2003).

Accelerated turnover rates of rhizosphere microbial biomass due to faunal grazing may be another explanation for the low microbial assimilation efficiency of root exudates as measured. Faunal grazing on rhizosphere bacteria and fungi has been suggested as a key factor of the "priming effect" of root exudates (Ingham *et al.* 1985; see Chapters 3 and 5). The high population density of bacteria in the rhizosphere (Foster 1988) may attract many grazers. The densities of both protozoa and bacterial-feeding nematodes have been shown to be higher in the rhizosphere than in the bulk soil (Ingham *et al.* 1985). Faunal grazing will increase the turnover rate of carbon and nitrogen in the rhizosphere, and subsequently result in a lower amount of exudate carbon or nitrogen in microbial biomass form, and higher amount being released as CO_2 (see Chapter 3). A rapid turnover rate (50% loss in 1 week) of microbial biomass-C formed from utilizing root exudates has been reported in a study with pine seedlings grown in soil using a pulse-labeling technique (Norton *et al.* 1990).

Another possible cause of the lower microbial assimilation efficiency of root exudates in the rhizosphere is the limitation of mineral nutrients such as nitrogen due to the competition with root uptake. Because of the abundant supply of available carbon in the form of exudates, microbial growth in the rhizosphere may be highly limited by mineral nutrients. For example, microbial respiration rate in the young wheat rhizosphere is not stimulated by addition of glucose (Cheng *et al.* 1994), but the assimilation efficiency of root-derived materials is higher when more nitrogen fertilizer has been used (Liljeroth *et al.* 1990; Van Veen *et al.* 1991).

The microbial assimilation efficiency of root exudates has deeper implications for carbon cycling in terrestrial ecosystems. It determines the flux of carbon entering the soil organic matter pool through living roots. As predicted global climatic change and the doubling of atmospheric CO_2 concentration in the near future may increase plant primary production and subsequently increase root exudate production (Kuikman *et al.* 1991), the microbial assimilation efficiency of root exudates will be a determinant as to what proportion of this increased primary production will be transferred to the soil organic carbon pool. If we assume that approximately 5 percent of the current global terrestrial primary production ($\sim 120 \times 10^{15}$ yr^{-1}) is in the form of root exudates, which equals to 6×10^{15} g C yr^{-1} (similar to the annual rate of current global fossil fuel consumption), and that a doubling of atmospheric CO_2 concentration will increase root exudate production by 70 percent (see Table 2.2), the amount of the CO_2-enhanced root exudate production would be 4.2×10^{15} g C yr^{-1}. If microbial utilization of exudates is complete (or 100% used), a range of microbial assimilation efficiencies of 5–20 percent will mean that the amount of the CO_2-enhanced carbon input into the soil via this route will vary between 0.21×10^{15} g C yr^{-1} and 0.84×10^{15} g C yr^{-1}, or from 3.5 to 14 percent of the annual rate of current global fossil fuel consumption.

2.6 TEMPORAL DYNAMICS OF EXUDATION AND RESPIRATION

The timing of root exudation determines how closely rhizosphere processes are linked with plant photosynthesis and aboveground physiology, and therefore the response time of rhizosphere activities to any change of environment aboveground. Pertaining to the temporal connections of root respiration, exudation, and rhizosphere microbial respiration of exudates, some pulse-labeling studies have reported contradictory results. Some studies seem to indicate that there exists a noticeable time lag between the time when root-respired ^{14}C-labeled CO_2 is released to the rhizosphere and the time of appearance of ^{14}C-labeled CO_2 from rhizo-microbial respiration of new root exudates (Warembourg and Billes 1979; Kuzyakov and Domanski 2002). This time lag may occur either between root respiration of the ^{14}C-labeled photosynthates and the appearance of the new rhizodeposits or between the time of rhizodeposit appearance in the rhizosphere and the time of microbial uptake and utilization, or both. Some other studies did not detect any meaningful time lag between these processes (Cheng *et al.* 1993, 1994). Understanding of the temporal aspect of these processes is required when dynamic models are used to simulate rhizosphere carbon fluxes (Kuzyakov *et al.* 1999; Luo *et al.* 2001). This assumed time lag has also been used to separate root respiration from rhizosphere microbial respiration (Kuzyakov *et al.* 1999, 2001).

    Using ^{14}C pulse-labeling techniques in a liquid culture of young wheat plants, Warembourg and Billes (1979) found that there was more than one distinctive peak of $^{14}CO_2$ release from the rhizosphere when microorganisms were present. The first peak of $^{14}CO_2$ release was assumed to be produced from root respiration, whereas microbial utilization of rhizodeposits was hypothesized responsible for the second peak, indicating that there was a time lag between the time when root-respired ^{14}C-labeled CO_2 is released to the rhizosphere and the time of appearance of ^{14}C-labeled CO_2 from rhizo-microbial respiration of new rhizodeposits. The time interval between the two peaks was roughly 24 hours. This time lag hypothesis was also supported by the fact that the occurrence of the second ^{14}C-labeled CO_2 peak is also coupled with an accumulation of ^{14}C-labeled rhizodeposits in the culturing solution of the non-sterile system. However, a small but visible second peak also appeared in the sterile treatment at a similar time interval, which could not be explained by the time lag hypothesis. Multiple peaks of $^{14}CO_2$ release from the rhizosphere were also reported by Nguyen et al. (1999) and Kuzyakov (2002). In a study of continuous release of rhizospheric CO_2 from 5-week-old maize plants, Nguyen et al. (1999) showed that there were two clearly distinguishable peaks of $^{14}CO_2$ release after the start of the ^{14}C-labeling, and that the time interval between the two peaks was 13.6 hours. This 13.6-hour interval between the two peaks did not correspond to a 24-hour diurnal cycle. Using a continuous liquid elution method, Kuzyakov and Siniakina (2001) chased rhizosphere release of both $^{14}CO_2$ and exudates for 4 days after a pulse labeling of *Lolium perenne* with $^{14}CO_2$. Their study showed that there were clearly two peaks of exudate release and a much smaller second peak of $^{14}CO_2$ release after the pulse labeling. The time interval between the first peak and the second peak roughly corresponded to the diurnal cycle. The authors believed that the dynamics of exudate release was mainly driven by photosynthate loading during the light period and the consumption of photo-assimilates during the dark period. The study also showed that the ^{14}C-labeled new photosynthate was simultaneously utilized in both root respiration and exudation processes, and that there was no detectable time lag between them. Their results could not render a clear answer to the question of whether or not there was a time lag between the time of rhizodeposit appearance in the rhizosphere and the time of microbial uptake and utilization, because root respiration could not be separately measured from rhizosphere microbial respiration in their experiment. The answer to this question could be found in one of our own studies (Cheng et al. 1993). In the study, root respiration and rhizosphere microbial respiration were separately measured for a short period of time after a pulse labeling by using an isotope trapping technique with the addition of ^{12}C-glucose solution. The addition of glucose solution reduced the release of $^{14}CO_2$ from the wheat rhizosphere by as much as 50 percent, primarily due to the reduction of rhizosphere microbial respiration because of substrate competition. This result clearly indicated that exudates were instantly utilized and converted into $^{14}CO_2$ by the rhizosphere microorganisms. No detectable time lag existed between the time of exudate appearance in the rhizosphere and the time of microbial uptake and utilization. This conclusion may also be inferred indirectly from the fact that adding sugar solution to soils commonly produces an instant pulse of CO_2 from microbial metabolism (Anderson and Domsch 1978). However, this may only apply to the case of readily available water-soluble exudates. It is probable that there exists a detectable time lag in hours or days either between root respiration of the ^{14}C-labeled photosynthates and the appearance of the new insoluble components of the rhizodeposits or between the time of appearance of the new insoluble components of the rhizodeposits and the time of microbial uptake and utilization of such components, or both. For example, microbial utilization of insoluble rhizodeposits may occur 2–5 days after the start of the pulse labeling (Warembourg and Billes 1979; Kuzyakov *et al.* 2001). Further studies are needed to advance our understanding on this issue.

In reconciliation of the above-mentioned results, we constructed a time-series model of carbon releases in the rhizosphere after a pulse labeling with $^{14}CO_2$ (Figure 2.2). The model depicts the following two points of understanding: (1) Current photo-assimilate production and translocation imposes the first level of control on the release of carbon sources from roots to the rhizosphere, because both exudate production and the efflux of rhizosphere CO_2 correspond to the diurnal cycle of photosynthesis; and (2) The occurrence

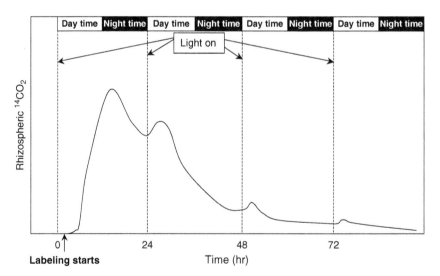

FIGURE 2.2   An idealized model of temporal dynamics of $^{14}CO_2$ released from the rhizosphere after a pulse labeling.

of more than one peaks of $^{14}CO_2$ release after a pulse labeling does not necessarily indicate that there exists a time lag either between root respiration of the ^{14}C-labeled photosynthates and the appearance of the new exudates or between the time of exudate appearance in the rhizosphere and the time of microbial uptake and utilization. As mentioned above, the time interval between the first peak and the second peak varies among different experiments, for example approximately 24 hours in Warembourg and Billes (1979), but 13.6 hours in Nguyen et al. (1999). The timing of the first peak is largely determined by the starting time and the duration of the labeling. However, the onset of the second peak and the third peak generally occur at the start of the light period (Todorovic et al. 2001). Therefore, different starting time and duration of the labeling in reference to the regular diurnal cycle give different time intervals between the initial two peaks among these experiments.

The discussion of the "time lag" issue is primarily based on data from pulse labeling studies of young herbaceous plants. The time dynamics of rhizosphere carbon release from woody plants awaits further investigation.

2.7 METHODS FOR STUDYING RHIZOSPHERE CARBON FLUXES

Many researchers may agree with the statement that method development has been, and remains, a key prerequisite for advancing rhizosphere science. Our understanding of carbon fluxes in the rhizosphere has significantly increased as new methods and approaches have been developed and used in rhizosphere research in the last several decades. Early studies utilized nutrient solution-based methods in order to estimate rates of rhizodeposition and assess the composition of root exudates (Whipps 1990). Various isotopes of carbon have been used to trace carbon pathways within the plant and through the plant–soil continuum (Nguyen 2003) using various labeling methodology. Recently, a series of molecular techniques have become available to evaluate the composition and identify sources of exudates (Killham and Yeomans 2001; Marschner 2003). However, the available methods have so far fallen short of providing accurate estimates of in situ rhizodeposition. Several factors contribute to this lack of data. A review of the literature provides a series of factors that are thought to influence rhizodeposition. These include root impedance (soil type, structure), nutrient status, pH, presence of microbial and faunal populations, temperature, light intensity, CO_2 concentration, stage of plant development, and presence of mycorrhizal associations (Whipps 1990; Killham and Yeomans 2001).

This complexity of potential interactions introduces questions of applicability of results of studies available to date for modeling of carbon movement in the rhizosphere in situ (Toal *et al.* 2000), especially considering that the vast majority of studies are performed in controlled conditions in the laboratory or in a greenhouse. Modeling is further complicated by the large variety of measurement units used for reporting results, which make comparisons between methods, as well as comparisons of different studies utilizing the same method, difficult. Likewise, the choice of organisms for the majority of the experiments introduces additional sources of bias into the resulting data, since most of the experiments use young annual plants, largely cereals. Cereals have been bred to allocate a larger portion of biomass aboveground, which skews carbon budgeting attempts. However, the methods available now have produced results that shed light onto rhizosphere carbon dynamics. The main advantages and disadvantages of five major kinds of methods were summarized in Table 2.3.

TABLE 2.3   Main Advantages and Disadvantages of Methods used in Studying C Fluxes in the Rhizosphere

Method	Benefits	Drawbacks
Nutrient Solution	Allows identification of exudate materials and sites of exudation.	Far removed from real conditions, does not allow quantification of carbon lost through respiration.
Pulse-Chase	Provides information on carbon pathways in relation to plant ecophysiology. The label is preferentially found in labile (non-structural) carbon pools.	Unable to provide balances for ecosystem carbon. Cannot distinguish between root respired C and C that results from microbial mineralization of root-derived carbon.
Continuous Labeling	Allows creation of carbon budgets. Label distributed homogeneously throughout the plant. Allows estimation of carbon flux through soil microbial biomass.	Expensive, cumbersome, does not distinguish between root respiration and rhizosphere decomposition of root-derived materials.
Natural Abundance	Relatively simple techniques, do not require use of radioactive materials, allow distinction between soil and plant carbon decomposition.	Due to high level of noise in the system only useful for distinguishing large differences.
Reporter Genes	Allows identification of spatial sources and compounds released into the rhizosphere.	Requires specialized equipment and training, does not provide data on carbon lost through respiration.

After Killham and Yeomans (2001), Whipps (1990).

Nutrient culture studies have provided a significant amount of information on the types of compounds exuded by plant roots into the rhizosphere, and allow differentiation of carbon lost as low molecular weight compounds from carbon lost as sloughed off root cells and root hairs (Nguyen 2003). However, inherent in the nutrient culture methodologies is the separation of the root–microbial complex of the rhizosphere, which has a high potential of disrupting the feedback mechanisms that may drive exudation. Likewise, resorption of exudates makes final estimates questionable. When physical barriers, such as sterile sand or glass beads, are introduced, the amount of exudates ranges tremendously; for instance, in *Hordeum vulgare* the range for exudates was $76–157 \, \mu g \, plant^{-1} \, day^{-1}$ (Barber and Gunn 1974); and in maize the addition of glass ballotini to mimic soil texture increased exudation from 94 to $280 \, mg \, g^{-1}$ dry weight of root over 5 days (Whipps 1990). Studies that rely on this technique often use seedlings, primarily of herbaceous plants, which may create additional bias. Lack of soil fauna and symbiotic organisms, which have the potential to influence rhizodeposition (see Chapters 3 and 4), also prohibitively limit the application of results from nutrient culture studies. Although nutrient culture studies have severe limitations for the application of the results to our understanding of in situ processes, recent advances in GC–MS analysis, based on the liquid culture technique, may allow a further understanding of the types of compounds exuded by plants, although quantitative information resulting from these methodologies remains suspect.

Pulse-chase studies involve exposing a plant to various isotopes of carbon for a short period of time, with subsequent evaluation of the sinks of this assimilated carbon both within the plant and in the rhizosphere. Pulse-chase studies largely provide information on C fluxes in relation to plant ecophysiology. Different lengths of labeling and chasing periods have been used in various experiments, which make comparison between different studies difficult. An analysis of 43 studies shows that exposure to the isotope ranges from 20 minutes to 720 hours with a mean of 6 and a median of 108 hours, with similar scales of variation in the duration and timing of the chase. However, a recent study indicates that the relative distribution of the tracer is not significantly influenced by the duration of the labeling, as long as the subsequent chasing period is long enough (Warembourg and Estelrich 2000). Due to the short exposure time in most experiments, the pulse-chase method cannot provide data for the construction of carbon budgets of the rhizosphere, nor does it allow partition between root and microbial respiration. Additional complications arise because carbon distribution determined at one plant development stage cannot be applied to others, which presents a problem since most experiments are done on very young plants (Kuzyakov and Domanski 2000).

Continuous labeling studies provide data that allow construction of complete carbon accounting models. However, most continuous labeling techniques are cumbersome, expensive, and not applicable in field situations. Continuous labeling studies are generally short term (mean 37, median 28 days). Methodological difficulties are numerous, for instance separation of very fine roots from soil is difficult, therefore some may be left in the sample, affecting the resulting data on carbon movement into soil.

A recent variation on the continuous labeling studies is the natural abundance method (e.g., Cheng 1996; Hanson *et al.* 2000), which uses naturally occurring isotopic composition differences to separate root- from soil-derived materials, and allows development of belowground carbon budgets without the expensive and difficult experimental setups, and does not require separation from the ambient atmosphere. However, several researchers have pointed out that this is a noisy system (Killham and Yeomans 2001), and therefore only large differences between treatments can be distinguished. Due to the fact that it uses unnatural plant–soil combinations, the applicability of the results obtained by this method to ecosystem carbon accounting may come under question.

Recent advances in molecular techniques allow tracing of low molecular weight compounds exuded by the roots in the rhizosphere, both providing spatial analysis of exudation sites and offering an assessment of the classes of compounds exuded by the roots (Killham and Yeomans 2001). However, these techniques require a set of very specific molecular tools and skills, as well as expensive equipment for genetic modification of microbial populations. Moreover, they do not provide data on carbon lost from the rhizosphere due to respiration.

In examining existing methods, and the reliability of the results obtained through their utilization, as well as during development of new methodologies, we need to recognize the level of abstraction from in situ conditions that the methods entail. Evaluating the wealth of studies that show how dramatically external factors can affect rhizosphere processes, method development should aim toward more precise replication of field conditions in the laboratory, and ideally the development of robust methodologies for in situ minimal disturbance investigations of rhizosphere processes.

2.8   PROSPECTS FOR FUTURE RESEARCH

In the past decades, some significant progress has been made in our pursuit of understanding rhizosphere C fluxes. For example, the strong top-down control of rhizosphere C fluxes by photosynthesis has been highlighted at several levels of resolution from tree plantations (e.g., Högberg *et al.* 2001), to small plots in a grassland (e.g., Craine *et al.* 1999), to well-controlled laboratory experiments (e.g., Kuzyakov and Cheng 2001). Initial understanding of the influence of elevated atmospheric CO_2 concentrations on rhizosphere C fluxes has been attended both in laboratory experiments and in field experiments (e.g., Cheng 1999; Luo *et al.* 2001). The natural abundance of ^{13}C has increasingly been used in studies of rhizosphere C fluxes, which may offer some new and complementary advantages, as compared to commonly used ^{14}C-labeling methods (e.g., Cheng *et al.* 2003). Methods and tools from molecular biology have been employed in studies of rhizosphere C fluxes (see Killham and Yeomans 2001). In perspective of future research on C fluxes in the rhizosphere, we consider the following as some of the key areas in rhizosphere research in the near future.

MUCH LESS IS KNOWN ABOUT CARBON FLUXES IN TREE RHIZOSPHERES

In the past few decades, research on carbon fluxes in the rhizosphere has been mostly restricted to cereal crops (Nguyen 2003). Little work has been done on carbon fluxes in tree rhizospheres. Forests have been identified as important processors of carbon (Houghton 1993). However, our lack of understanding of below-ground carbon fluxes in forest systems, and specifically the role of roots, is the greatest limitation in our ability to assess the contribution of forests as global carbon processors (e.g., Schimel 1995). According to current estimates, total rhizosphere respiration may contribute, on average, approximately 50 percent of the total CO_2 released from belowground components in forest ecosystems (Hanson *et al.* 2000), and ranges from 30 to as high as 90 percent (Bowden *et al.* 1993). Therefore, carbon fluxes in the rhizosphere of forest ecosystems represent important belowground processes responsible for C release. However, the controlling mechanisms and the functional role of this large carbon expenditure are not well understood. It is commonly known that temperature strongly regulates fine root respiration in an exponential fashion (Ryan *et al.* 1996). Several studies have demonstrated that rhizosphere respiration is tightly coupled with photosynthesis in annual plants (Kuzyakov and Cheng 2001) with very short time lags (minutes to hours). As shown in a study by Horwath *et al.* (1994) using ^{14}C pulse labeling, time lags in the linkage between aboveground photosynthesis and rhizosphere respiration can be as short as a few days for very young hybrid poplars. The time lag has been reported to be in the range of 7–60 days in a study using ^{13}C signal from the FACE treatment at a loblolly pine site (Luo *et al.* 2001). As shown in a large-scale tree girdling experiment with a boreal Scots pine forest, the reduction in total soil CO_2 efflux caused by the termination of current photoassimilate supply to the roots system can be as high as 37 percent within

5 days, and 54 percent within 1–2 months (Högberg *et al.* 2001). However, the underlying mechanisms responsible for the coupling between photosynthesis and rhizosphere respiration remain largely unknown.

LINKING C DYNAMICS IN THE RHIZOSPHERE TO GENERAL MODELS OF C ALLOCATION

Although the introduction of various isotope tracer methods has lead to meaningful progress in assessing the quantity of rhizosphere C fluxes, less work has been done in conjunction with the framework of overall C allocation beyond a common practice of expressing C flux as a percentage of gross or net primary production. However, it is crucial to understand C fluxes in the rhizosphere in the context of mechanistic relations with plant C allocation and its controls, if one main purpose of the research is to scale the results to be generally applicable to a larger system or other kinds of systems. Most research so far has provided data either on the quantity of C fluxes in the rhizosphere for a particular system or on the qualitative influence of some environmental factors, such as lighting, elevated CO_2, grazing, and presence or absence of microorganisms. For these data to be scalable in a general model of C fluxes in the rhizosphere, we need to understand the quantitative formulation between a controlling factor of either biological or environmental nature and the relative change in the quantity of a C flux in response to the change of the ecological factor. A so-called "shared-control" hypothesis has been proposed to be generally applicable to the case of C fluxes in the rhizosphere by Farrar *et al.* (2003). This hypothesis stipulates that every step in the flow of C from photosynthesis to the final utilization in the rhizosphere contributes to the control of the overall flux. Based on results, primarily from liquid culture experiments, they suggested that photosynthesis exerts the bulk of the control on the size of C flux into the rhizosphere, and that active or passive exudation controls the C outflow more than microbial utilization. However, quantitative models describing such shared controls of C fluxes in the rhizosphere are still lacking because of the complexity involved in such multistep modeling exercises (Toal *et al.* 2000). Likewise, more complex studies involving experimental setups that closely mimic in situ conditions may further assist in developing a more sophisticated understanding of the controls exerted on the flux of C by the rhizosphere.

CHEMICAL COMPOSITION OF RHIZODEPOSITS IN SOILS

As we mentioned earlier, our understanding of the chemical composition of rhizodeposits is still entirely based on data from liquid culture experiments. We know very little about the chemical composition of rhizodeposits in soils before transformations by rhizosphere microorganisms, given that the understanding of the chemical nature of rhizodeposits is a prerequisite for a better handle on the ecological functions of rhizodeposits. Hopefully, the dual use of isotope tracers with reporter genes may offer new opportunities in this area of research (Killham and Yeomans 2001).

CARBON FLUXES AND THE COEVOLUTION BETWEEN PLANTS AND SOIL BIOTA

If higher plants invest a significant amount of fixed carbon into the rhizosphere to support a portion of the soil biota, the coevolution between plants and rhizosphere biota must shape the quantity and the quality of rhizosphere C fluxes through selection and adaptation. Given the known wide range of association types between roots and rhizosphere biota from highly mutualistic (e.g., rhizobium–legume) to totally opportunistic (e.g., free-living bacteria), we know little about how these different types of associations operate in determining the amount and the types of rhizosphere C flows, not to mention the potential role of the complex interactions in the rhizosphere through evolutionary time. This complexity and the associated opportunities for future research are well illustrated in the case of "the free rider" problem in the recent paper by Denison *et al.* (2003). The "free rider" problem arises when considering microbial intra-species competition in the context of plant–microbe cooperation and mutualisms. Because the benefits (e.g., carbon substrates) gained from an individual plant via microbial cooperation are often used by many individual microbes, those microbial individuals that do not provide the cost of the cooperation (e.g., N_2-fixation) yet may equally gain the benefits (often in the form of mutants), or the "free-riders," should have the tendency to replace those microbial individuals that do bear the cost of such cooperation. Yet, this reasoning directly contradicts the fact that plant–microbe cooperation and mutualisms have persisted, presumably, for millions of years. Using this apparent paradox as a thread, Denison *et al.* (2003) discussed potential mechanisms of plant–microbe cooperation and mutualisms and touched on many intricate connections in the rhizosphere.

ACKNOWLEDGEMENTS

We thank the help from an anonymous reviewer on an earlier draft. The work of this chapter was supported by National Research Initiative Competitive Grant no. 2003-35107-13716 from the USDA Cooperative State Research, Education, and Extension Service, and a grant from the Kearney Foundation of Soil Science.

REFERENCES

Abbadie, L., A. Mariotti, and J. Menaut. 1992. Independence of savanna grass from soil organic matter for their nitrogen supply. *Ecology* 73:608–613.

Anderson, J.P.E., and K.H. Domsch. 1978. A physiological method for the quantitative measurement of microbial biomass in soils. *Soil Biology and Biochemistry* 10:215–221.

Arnone, J.A., and J.C. Gordon. 1990. Effect of nodulation, nitrogen fixation and CO_2 enrichment on the physiology, growth and dry mass allocation of seedlings of *Alnus rubra* Bong. *New Phytologist* 116:55–66.

Barber, D.A., and K.B. Gunn. 1974. The effect of mechanical forces on the exudation of organic substances by the roots of cereal plants grown under sterile conditions. *New Phytologist* 73:69–80.

Bertin, C., X. Yang, and L. Weston. 2003. The role of root exudates and allelochemicals in the rhizosphere. *Plant and Soil* 256(1):67–83.

Billes, G., H. Rouhier, and P. Bottner. 1993. Modifications of the carbon and nitrogen allocations in the plant (*Triticum aestivum* L.) soil system in response to increased atmospheric CO_2 concentration. *Plant and Soil* 157:215–225.

Bowden, R.D., K.J. Nadelhoffer, R.D. Boone, J.M. Melillo, and J.B. Garris. 1993. Contributions of aboveground litter, belowground litter, and root respiration to total soil respiration in a temperate mixed hardwood forest. *Canadian Journal of Forest Research* 23:1402–1407.

Cardon, Z.G., A.D. Czaja, J.L. Funk, and P.L. Vitt. 2002. Periodic carbon flushing to roots of *Quercus rubra* saplings affects soil respiration and rhizosphere microbial biomass. *Oecologia* 133:215–223.

Cheng, W. 1996. Measurement of rhizosphere respiration and organic matter decomposition using natural ^{13}C. *Plant and Soil* 183:263–268.

Cheng, W. 1999. Rhizosphere feedbacks in elevated CO_2. *Tree Physiology* 19:313–320.

Cheng, W., and D.C. Coleman. 1990. Effect of living roots on soil organic matter decomposition. *Soil Biology and Biochemistry* 22:781–787.

Cheng, W., and D.W. Johnson. 1998. Effect of elevated CO_2 on rhizosphere processes and soil organic matter decomposition. *Plant and Soil* 202:167–174.

Cheng, W., and Kuzyakov, Y. 2005. Root effects on soil organic matter decomposition. In Zobel, R.W., and Wright, S.F. (eds) *Roots and Soil Management: Interactions Between Roots and the Soil*, Agronomy Monograph no. 48, ASA-CSSA-SSSA, Madison, Wisconsin, USA.

Cheng, W., D.C. Coleman, C.R. Carroll, and C.A. Hoffman. 1993. *In situ* measurement of root respiration and soluble carbon concentrations in the rhizosphere. *Soil Biology and Biochemistry* 25:1189–1196.

Cheng, W., D.C. Coleman, C.R. Carroll, and C.A. Hoffman. 1994. Investigating short-term carbon flows in the rhizospheres of different plant species using isotopic trapping. *Agronomy Journal* 86:782–791.

Cheng, W., D.A. Sims, Y. Luo, D.W. Johnson, J.T. Ball, and J.S. Coleman. 2000. Carbon budgeting in plant-soil mesocosms under elevated CO_2: Locally missing carbon? *Global Change Biology* 6:99–110.

Cheng, W., D.W. Johnson, and S. Fu. 2003. Rhizosphere effects on decomposition: controls of plant species, phenology, and fertilization. *Soil Science Society of America Journal* 67:1418–1427.

Craine, J.M., D.A. Wedin, and F.S. Chapin. 1999. Predominance of ecophysiological controls on soil CO_2 flux in a Minnesota grassland. *Plant and Soil* 207:77–86.

Denison, R.F., C. Bledsoe, M. Kahn, F.O. Gaa, E.L. Simms, and L.S. Thomashow. 2003. Cooperation in the rhizosphere and the "free rider" problem. *Ecology* 84:838–845.

Dyer, M.I., M.A. Acra, G.M. Wang, D.C. Coleman, D.W. Freckman, S.J. McNaughton, and B.R. Strain. 1991. Source-sink carbon relations in two *Panicum coloratum* ecotypes in response to herbivory. *Ecology* 72:1472–1483.

Eissenstat, D.M., and R.D. Yanai. 1997. The ecology of root lifespan. *Advances in Ecological Research* 27:1–60.

Elliott, E.T., C.V. Cole, B.C. Fairbanks, L.E. Woods, R.J. Bryant, and D.C. Coleman. 1983. Short-term bacterial growth, nutrient uptake, and ATP turnover in sterilized, inoculated and C-amended soil: The influence of N availability. *Soil Biology and Biochemistry* 15:85–91.

Farrar, J.F., and D.L. Jones. 2000. The control of carbon acquisition by roots. *New Phytologist* 147:43–53.

Farrar, J.F., M. Hawes, D. Jones, and S. Lindow. 2003. How roots control the flux of carbon to the rhizosphere. *Ecology* 84:827–837.

Foster, R.C. 1988. Microenvironments of soil microorganisms. *Biology and Fertility of Soils* 6:189–203.

Fu, S., and W. Cheng. 2002. Rhizosphere priming effects on the decomposition of soil organic matter in C_4 and C_3 grassland soils. *Plant and Soil* 238:289–294.

Garrett, C.J., D.A. Crossley, D.C. Coleman, P.F. Hendrix, K.W. Kisselle, and R.L. Porter. 2001. Impact of the rhizosphere on soil microarthropods in agroecosystems on the Georgia piedmont. *Applied Soil Ecology* 16:141–148.

Hamilton, E.W., and D.A. Frank. 2001. Can plant stimulate soil microbes and their own nutrient supply? Evidence from a grazing tolerant grass. *Ecology* 82:2397–2402.

Hanson, P.J., N.T. Edwards, C.T. Garten, and J.A. Andrews. 2000. Separating root and soil microbial contributions to soil respiration: A review of methods and observations. *Biogeochemistry* 48:115–146.

Harris, D., and E.A. Paul. 1991. Techniques for examining the carbon relationships of plant-microbial symbioses. In: Coleman, D.C., and Fry, B. (eds) *Carbon Isotope Techniques*. Academic Press, San Diego, CA, pp. 39–52.

Helal, H.M., and D. Sauerbeck. 1989. Carbon turnover in the rhizosphere. *Zeitschrift fur Pflanzenernahrung und Bodenkunde* 152:211–216.

Herald, A. 1980. Regulation of photosynthesis by sink activity-the missing link. *New Phytologist* 86:131–144.

Hirsch, A.M., W.D. Bauer, D.M. Bird, J. Cullimore, B. Tyler, and J. Yoder. 2003. Molecular signals and receptors: Controlling rhizosphere interactions between plants and other organisms. *Ecology* 84:858–868.

Högberg, P., A. Nordgren, N. Buchmann, A.F.S. Taylor, A. Ekblad, M.N. Hogberg, G. Nyberg, M. Ottosson-Lofvenius, and D.J. Read. 2001. Large-scale forest girdling shows that current photosynthesis drives soil respiration. *Nature* 411:789–792.

Horwath, W.R., K.S. Pregitzer, and E.A. Paul. 1994. C-14 allocation in tree soil systems. *Tree Physiology* 14:1163–1176.

Houghton, R.A. 1993. Is carbon accumulating in the northern temperate zone? *Global Biogeochemical Cycles* 7:611–617.

Hungate, B.A., E.A. Holland, R.B. Jackson, F.S. Chapin, H.A. Mooney, and C.B. Field. 1997. The fate of carbon in grasslands under carbon dioxide enrichment. *Nature* 388:576–579.

Ingham, R.E., J.A. Trofymow, E.R. Ingham, and D.C. Coleman. 1985. Interactions of bacteria, fungi, and their nematode grazers: Effects of nutrient cycling and plant growth. *Ecological Monographs* 55:119–140.

Jaeger, C.H. III, S.E. Lindow, W. Miller, E. Clark, and M.K. Firestone. 1999. Mapping of sugar and amino acid availability in soil around roots with bacterial sensors of sucrose and tryptophan. *Applied and Environmental Microbiology* 65:2685–2690.

Johansson, G. 1992. Release of organic C from growing roots of meadow fescue (*Festuca pratensis* L.). *Soil Biology and Biochemistry* 24:427–433.

Jones, D.L., and P.R. Darrah. 1995. Influx and efflux of organic acids across the soil-root interface of *Zea mays* L. and its implications in rhizosphere C flow. *Plant and Soil* 173:103–109.

Jones, D.L., J. Farrar, and K.E. Giller. 2003. Associative nitrogen fixation and root exudation–What is theoretically possible in the rhizosphere? *Symbiosis* 35:19–28.

Kapulnik, Y. 1991. Nonsymbiotic nitrogen-fixing microorganisms. In: Waisel, Y., Eshel, A., and Kafkafi, U. (eds) *Plant Roots: The Hidden Half.* Marcel Dekker, Inc., New York, pp. 703–716.

Killham, K., and C. Yeomans. 2001. Rhizosphere carbon flow measurement and implications: from isotopes to reporter genes. *Plant and Soil* 232:91–96.

Kuikman, P.J., L.J.A. Lekkerkerk, and J.A. van Veen. 1991. Carbon dynamics of a soil planted with wheat under elevated CO_2 concentration. In: Wilson, W.S. (ed.) *Advances in Soil Organic Matter Research: The Impact on Agriculture and the Environment, vol Special Publication 90.* The Royal Society of Chemistry, Cambridge, UK, pp. 267–274.

Kuzyakov, Y. 2002. Separating microbial respiration of exudates from root respiration in non-sterile soils: A comparison of four methods. *Soil Biology and Biochemistry* 34:1621–1631.

Kuzyakov, Y., and W. Cheng. 2001. Photosynthesis controls of rhizosphere respiration and organic matter decomposition. *Soil Biology and Biochemistry* 33:1915–1925.

Kuzyakov, Y., and G. Domanski. 2000. Carbon inputs by plants into the soil. Review. *Journal of Plant Nutrition and Soil Science* 163:421–431.

Kuzyakov, Y., and G. Domanski. 2002. Model for rhizodeposition and CO_2 efflux from planted soil and its validation by C-14 pulse labeling of ryegrass. *Plant and Soil* 239:87–102.

Kuzyakov, Y., and S.V. Siniakina. 2001. A novel method for separating root-derived organic compounds from root respiration in non-sterile soils. *Journal of Plant Nutrition and Soil Science* 164:511–517.

Kuzyakov, Y., A. Kretzschmar, and K. Stahr. 1999. Contribution of *Lolium perenne* rhizodeposition to carbon turnover of pasture soil. *Plant and Soil* 213:127–136.

Kuzyakov, Y., H. Ehrensberger, and K. Stahr. 2001. Carbon partitioning and below-ground translocation by *Lolium perenne. Soil Biology and Biochemistry* 33:61–74.

Lambers, H. 1987. Growth, respiration, exudation and symbiotic associations: The fate of carbon translocated to the roots. In: Gregory, P.J., and Lake, J.V. (eds) *Root Development and Function.* Cambridge University Press, London, pp. 125–145.

Lekkerkerk, L.J.A., S.C. van De Geijn, and J.A. van Veen. 1990. Effects of elevated atmospheric CO_2-levels on the carbon economy of a soil planted with wheat. In: Bouwman, A.F. (ed.) *Soils and the Greenhouse Effect.* John Wiley & Sons, pp. 423–429.

Liljeroth, E., J.A. Van Veen, and H.J. Miller. 1990. Assimilate translocation to the rhizosphere of two wheat lines and subsequent utilization by rhizosphere microorganisms at two nitrogen concentrations. *Soil Biology and Biochemistry* 22:1015–1021.

Luo, Y.Q., L.H. Wu, J.A. Andrew, L. White, R. Matamala, K.W.R. Schafer, and W.H. Schlesinger. 2001. Elevated CO_2 differentiates ecosystem carbon processes: Deconvolution analysis of Duke FACE data. *Ecological Monographs* 71:357–376.

Marschner, H. 1995. *Mineral Nutrition of Higher Plants.* Academic Press, San Diego.

Marschner, P. 2003. BIOLOG and Molecular Methods to Assess Root-Microbe Interactions in the Rhizosphere. *Third International Symposium on Dynamics of Physiological Processes in Woody Roots. School of Plant Biology.* University of Western Australia, Perth, Australia, p. 41.

Martin, J.K., and R. Merckx. 1992. The partitioning of photosynthetically fixed carbon within the rhizosphere of mature wheat. *Soil Biology and Biochemistry* 24:1147–1156.

Matamala, R., and W.H. Schlesinger. 2000. Effects of elevated atmospheric CO_2 on fine root production and activity in an intact temperate forest ecosystem. *Global Change Biology* 6:967–979.

Nguyen, C. 2003. Rhizodeposition of organic C by plants: Mechanisms and controls. *Agronomie* 23:375–396.

Nguyen, C., C. Todorovic, C. Robin, A. Christophe, and A. Guckert. 1999. Continuous monitoring of rhizosphere respiration after labeling of plant shoots with $^{14}CO_2$. *Plant and Soil* 212:191–201.

Norby, R.J., E.G. O'Neill, and R.J. Luxmoore. 1986. Effects of CO_2 enrichment on the growth and mineral nutrition of *Quercus alba* seedlings in nutrient-poor soil. *Plant Physiology* 82:83–89.

Norby, R.J., P.J. Hanson, E.G. O'Neill, T.J. Tschaplinski, J.F. Weltzin, R.A. Hansen, W.X. Cheng, S.D. Wullschleger, C.A. Gunderson, N.T. Edwards, and D.W. Johnson. 2002. Net primary productivity of a CO_2-enriched deciduous forest and the implications for carbon storage. *Ecological Applications* 12:1261–1266.

Norton, J.M., J.L. Smith, and M.K. Firestone. 1990. Carbon flow in the rhizosphere of ponderosa pine seedlings. *Soil Biology and Biochemistry* 22:449–455.

Paterson, E., E.A.S. Rattray, and K. Killham. 1996. Effects of elevated CO_2 concentration on C-partitioning and rhizosphere C-flow for three plant species. *Soil Biology and Biochemistry* 28:195–201.

Paterson, E., A. Hodge, B. Thornton, P. Millard, and K. Killham. 1999. Carbon partitioning and rhizosphere C-flow in *Lolium perenne* as affected by CO_2 concentration, irridiance and belowground conditions. *Global Change Biology* 5:669–678.

Payne, W.J. 1970. Energy yields and growth of heterotrophs. *Annual Review of Microbiology* 24:17–52.

Phillips, D.A., H. Ferris, D.R. Cook, and D.R. Strong. 2003. Molecular control points in rhizosphere food webs. *Ecology* 84:816–826.

Post, W.M., T.H. Peng, W.R. Emmanuel, A.W. King, V.H. Dale, and D.L. DeAngelis. 1990. The global carbon cycle. *American Scientist* 78:310–326.

Rouhier, H., G. Billes, L. Billes, and P. Bottner. 1996. Carbon fluxes in the rhizosphere of sweet chestnut seedlings (*Castanea sativa*) grown under two atmospheric CO_2 concentrations: ^{14}C partitioning after pulse labelling. *Plant and Soil* 180:101–111.

Ryan, M.G., R.M. Hubbard, S. Pongracic, R.J. Raison, and R.E. McMuurtrie. 1996. Foliage, fine-root, woody tissue and stand respiration in *Pinus radiata* in relation to nitrogen status. *Tree Physiology* 16:333–343.

Schimel, D.S. 1995. Terrestrial ecosystems and the carbon cycle. *Global Change Biology* 1:77–91.

Thomas, R.B., D.D. Richter, H. Ye, P.R. Heine, and B.R. Strain. 1991. Nitrogen dynamics and growth of seedlings of an N-fixing tree (*Gliricidia sepium*) exposed to elevated atmospheric carbon dioxide. *Oecologia* 8:415–421.

Tissue, D.T., J.P. Megonigal, and R.B. Thomas. 1997. Nitrogenase activity and N_2 fixation are stimulated by elevated CO_2 in a tropical N_2-fixing tree. *Oecologia* 109:28–33.

Toal, M.E., C. Yeomans, K. Killham, and A.A. Meharg. 2000. A review of rhizosphere carbon flow modeling. *Plant and Soil* 222:263–281.

Todorovic, C., C. Nguyen, C. Robin, and A. Guckert. 2001. Root and microbial involvement in the kinetics of C-14-partitioning to rhizosphere respiration after a pulse labeling of maize assimilates. *Plant and Soil* 228:179–189.

Van Veen, J.A., E. Liljeroth, L.J.A. Lekkerkerk, and S.C. Van de Geijn. 1991. Carbon fluxes in plant-soil systems at elevated atmospheric CO_2 levels. *Ecological Applications* 1:175–181.

Warembourg, F.R., and G. Billes. 1979. Estimating carbon transfers in the plant rhizosphere. In: Harley, J.L., and Russell, R.S. (eds) *The Soil Root Interface*. Academic Press, London, UK, pp. 182–196.

Warembourg, F.R., and H.D. Estelrich. 2000. Towards a better understanding of carbon flow in the rhizosphere: A time-dependent approach using carbon-14. *Biology and Fertility of Soils* 30:528–534.

Warembourg, F.R., C. Roumet, and F. Lafont. 2003. Differences in rhizosphere carbon-partitioning among plant species of different families. *Plant and Soil* 256:347–357.

Whipps, J.M. 1990. Carbon Economy. In: Lynch, J.M. (ed.) *The Rhizosphere*. John Wiley & Sons, New York.

Whipps, J.M. 2001. Microbial interactions and biocontrol in the rhizosphere. *Journal of Experimental Botany* 52:487–511.

Whipps, J.M., and J.M. Lynch. 1985. Energy losses by the plant in rhizodeposition. *Annual Proceedings of the Phytochemical Society of Europe* 26:59–71.

Zak, D.R., and K.S. Pregitzer. 1998. Integration of ecophysiological and biogeochemical approaches to ecosystem dynamics. In: Groffman, P.M. (ed.) *Successes, Limitations, and Frontiers in Ecosystem Science*. Springer, New York, USA, pp. 372–403.

CHAPTER 3

Microfaunal Interactions in the Rhizosphere, How Nematodes and Protozoa Link Above- and Belowground Processes

Bryan S. Griffiths, Søren Christensen, and Michael Bonkowski

3.1 THE PLANT AS A BRIDGE BETWEEN ABOVE- AND BELOWGROUND POPULATIONS AND PROCESSES

The living plant is the basis of existence for several groups of organisms both above- and belowground. These organisms include, for example, mycorrhizal fungi, free-living rhizosphere organisms, foliar and root herbivorous insects, root pathogenic fungi, and nematodes. These organisms affect plant growth, and are affected by plant growth, but we currently have an incomplete understanding of the cumulative and interactive effects of all these organisms since our knowledge is mainly based on isolated investigations of single organism groups. To further our understanding we need to know how these consumer organisms develop and interact with one another during plant growth (i.e., the above- and belowground multitrophic interactions *sensu* van der Putten *et al.* 2001). Should the plant be regarded solely as a supplier of carbon to the various consumers, or as an organism that interacts with its consumers forming a bridge between above- and belowground populations and processes? In this chapter we bring together our knowledge of microfauna (nematodes and protozoa) in the rhizosphere with the latest experimental findings on plant–microbe interactions to conclude that the connection between plants and soil animals is far more than simply that of resource and consumer.

3.2   RHIZOSPHERE MICROFAUNA – DIRECT EFFECTS ON CARBON AND NITROGEN FLOWS

Previous reviews of the role of microfauna (nematodes and protozoa) in the rhizosphere have tended to concentrate on their contribution to gross flows of carbon and nitrogen (see, for example, Griffiths 1994; Zwart *et al.* 1994) or their role in disease suppression (Curl and Harper 1990). In the rhizosphere, bacteria are more important decomposer organisms than fungi, because of the large supply of easily decomposable organic matter (Wardle 2002), so interactions between bacteria and their microfaunal grazers are of more consequence than those of fungi and their grazing animals. The activity of microorganisms in soil is generally limited by carbon, but not in the rhizosphere where plants steadily supply microorganisms with easily available carbon. Consequently, a specialized microflora typically consisting of fast-growing bacteria results in increased levels of microbial biomass and activity around plant roots (Alphei *et al.* 1996). There is strong top-down control of these bacterial populations by the grazing pressure of microbivorous nematodes and protozoa (Ingham *et al.* 1986; Wardle 2002). The release of carbon in the form of root exudates may account for up to 40 percent of the dry matter produced by plants (Lynch and Whipps 1990), and this topic is dealt within greater detail in Chapter 3. Even if the C-transfer to exudation was 10–20 percent of total net fixed carbon (Rovira 1991), other microbial symbionts such as mycorrhizae (Smith and Read 1997) or N_2-fixing microorganisms (Ryle *et al.* 1979) may each consume another 10–20 percent of total net fixed carbon, so that plants would still release up to half of their total fixed carbon to fuel microbial interactions in the rhizosphere.

We take the view that supporting microbial interactions in the rhizosphere must be of fundamental importance for plants to justify this significant input of carbon, which could otherwise be used to construct light-capturing or defensive structural tissues aboveground. In particular, why are plants providing ample energy in the form of exudates to a microbial community that is strongly competing with roots for available nutrients? The answer partly lies in the loop structure of the bacterial energy channel in the rhizosphere. Nutrients become only temporarily locked up in bacterial biomass near the root surface and are successively liberated by microfaunal grazing (Bonkowski *et al.* 2000a). The interplay between microorganisms and microfauna determines the rate of nutrient cycling and strongly enhances the availability of mineral nutrients to plants (Clarholm 1985; Ingham *et al.* 1985; Jentschke *et al.* 1995; Alphei *et al.* 1996; Bonkowski *et al.* 2000b, 2001b). The assumed mechanism, known as the "microbial loop in soil" (Clarholm 1985), is triggered by the release of root exudates from plants which increase bacterial growth in the rhizosphere. Microfaunal grazing in the rhizosphere is particularly important because plant-available nutrients will be strongly sequestered during microbial growth (Kaye and Hart 1997; Wang and Bakken 1997) and would remain locked up in bacterial biomass if consumption by nematodes and protozoa would not constantly remobilize essential nutrients for plant uptake (Bonkowski et al. 2000b).

The subsequent increase in plant N uptake is well documented in experimental systems (see previous references plus Ingham et al. 1985; Verhagen et al. 1995). Root-derived carbon leads to a general increase in the populations of microfauna in the rhizosphere, compared to bulk soil, up to 27-fold for free-living nematodes (Griffiths 1990) and 35-fold for protozoa (Zwart et al. 1994). However, the quantifiable benefit of this direct microfaunal activity to the gross N nutrition of plants in the field is slight. Nitrogen balance models indicate that this activity is only sufficient to allow for recycling of the N lost from the plant by exudation rather than to mineralize N from soil organic matter, and could only supply a small proportion of the measured uptake rates of N (Griffiths and Robinson 1992).

Microfaunal populations in the rhizosphere often reach a maximum on the older portions of the root system rather than the root-tip itself. On barley roots nematode populations reached a maximum on roots that were 10 days old (Griffiths et al. 1991), although large numbers of active amoebae occurred near the root-tip of plants growing on agar (Coûteaux et al. 1988; Bonkowski and Brandt 2002). The likely maximum effects of the microfauna can, therefore, be spatially (and temporally) removed from the location of exudation at the root-tip. The observation that bacterial populations oscillate as a kind of moving wave along a root (Semenov et al. 1999) maybe related to predator–prey dynamics induced by the grazing of microfauna on bacteria, although this was considered unlikely by these authors. Increases in rhizosphere protozoa occur mainly early in the life of an annual plant, during the nutrient acquisition phase before flowering (Rønn et al. 2002), further emphasizing temporal dynamics.

Microfauna do have significant, direct effects on rhizosphere C- and N-flow through the action of plant-parasitic nematodes. Low amounts of root infestation (typical of natural field densities) by the clover cyst nematode (Heterodera trifolii) on white clover (Trifolium repens) increased the translocation of photoassimilate to the roots, increased leakage of carbon from the roots and increased microbial biomass in the rhizosphere (Yeates et al. 1998). This increased flow of C was confirmed with subsequent studies on a further four species of root-feeding nematodes (Yeates et al. 1999). While clover root production was increased in response to low levels of infestation by clover cyst nematode, root biomass of a companion species not attacked by the nematode, perennial ryegrass (Lolium perenne), was also increased (Bardgett et al. 1999). This was due to increased fluxes of N from the clover being recycled and taken up by the ryegrass.

The direct effects of microbial-feeding nematodes and protozoa in the rhizosphere are to effectively recycle nutrients that would otherwise remain immobilized in the microbial biomass. Quantitatively, however, this only liberates as much N as the plant would lose through root exudation and does not represent a significant flow of N to the growing plant. The effects of low levels of root herbivory by plant-feeding nematodes directly contribute to rhizosphere C-flows and can also, in some circumstances, enhance N-availability.

3.3   RHIZOSPHERE MICROFAUNA – INDIRECT EFFECTS ON PLANT GROWTH

Indirect interactions of microfaunal grazing seem even more important than direct effects due to nutrient release (Bonkowski and Brandt 2002). Protozoa have, for example, been found to increase plant biomass independently of nutrient contents in the plant tissue (Alphei *et al.* 1996). Thus, in a laboratory experiment with a constant supply of excess nutrients, protozoa increased the biomass of spruce (*Picea abies*) seedlings up to 60 percent (Jentschke *et al.* 1995; Table 3.1).

Plants are not simply passive recipients of nutrients, but information from the environment affects their belowground allocations such as root proliferation (Hodge *et al.* 1999), formation of symbiotic relationships (e.g. mycorrhizal fungi, Smith and Read 1997; or N_2-fixing bacteria, Ryle *et al.* 1979), alteration in exudation rates (Bonkowski *et al.* 2001b; Wamberg *et al.* 2003), interactions with free-living bacteria (Joseph and Phillips 2003, Mathesius *et al.* 2003), or production of secondary defence compounds against herbivores (Cipollini *et al.* 2003). Since root morphology is both genetically programmed and environmentally determined (Rolfe *et al.* 1997), there must be signal transduction pathways that interpret complex environmental conditions and activate genes to enter a particular symbiosis or to form a lateral

TABLE 3.1   The Biomass and Root Development of Spruce (*Picea abies*) Seedlings, Grown in the Presence or Absence of Protozoa (Either a Natural Community of Soil Protozoa or a Limited Number of Species from Laboratory Culture), but with a Non-Limiting Supply of Nutrients. Means Followed by Different Letters are Significantly Different ($P < 0.05$), Data from Jentschke *et al.* 1995

	No protozoa	Mixed soil protozoa	Protozoa from culture
Shoot wt (mg)	278[a]	493[b]	406[b]
Root wt (mg)	401	571	572
Root tips (000's)	4.3[b]	13.3[a]	9.8[a,b]
Root length (m)	11[b]	31[a]	26[a]

root at a particular time and place, for example. The exchange of signals between plants and microorganisms is reciprocal and in case of root-infecting plant symbionts and pathogens an area of intense research (McKenzie Bird and Koltai 2000). Recently, Phillips and Strong (2003) introduced the concept of "rhizosphere control points" to emphasize the importance of information exchange between plants and microorganisms.

From a microbial perspective, the evolution of strategies capable of enhancing energy transfer to the roots would lead to a strong increase in fitness of those microorganisms that influence gene regulation in plants by sending the right signals. Specialized bacteria are the dominant colonizers of plant roots (Marilley and Aragno 1999) and indeed many of the rhizosphere bacteria have the potential to affect plant performance by producing hormones (Lambrecht *et al.* 2000). Up to 80 percent of the bacteria isolated from plant rhizospheres are considered to produce auxins (Patten and Glick 1996), and their widespread ability to produce cytokinins led Holland (1997) to suggest that cytokinins in plants may originate exclusively from microorganisms. The widespread ability of both beneficial and deleterious rhizosphere microorganisms to produce plant hormones suggests that rhizosphere bacteria play an important role in manipulating root and plant growth (Shishido *et al.* 1996; Rolfe *et al.* 1997). Recent molecular evidence points to the role of phytohormones in the induction of giant cells by root-knot nematodes (*Meloidogyne* spp.), and that the nematodes acquire the genes to synthesize or modulate phytohormones by horizontal gene transfer (McKenzie Bird and Koltai 2000).

The strong top-down regulation of rhizosphere bacteria by grazing gives a central role to nematodes and protozoa in the interactions between plant roots and their colonizing microorganisms. Such food web interactions are explored in greater detail in Chapter 12. Protozoa seem quite selective in their bacterial food choice (Boenigk and Arndt 2002) and significant changes in bacterial community composition due to protozoan grazing have been confirmed in freshwater systems (Posch *et al.* 1999) as well as in the rhizosphere of plants (Griffiths *et al.* 1999; Bonkowski and Brandt 2002). These grazing-induced changes in microbial composition affect fundamental ecosystem properties because soil bacteria occupy some of the most important control points for nutrient cycling and plant growth. For instance, N_2-fixing, nitrifying, and denitrifying bacteria dominate the nitrogen cycle (Mengel 1996). A strong stimulation of nitrifying bacteria is commonly observed in the presence of protozoan grazers, presumably through predation on their faster-growing bacterial competitors, resulting in high concentrations of NO_3^- in culture liquid and rhizosphere soil (Verhagen *et al.* 1995; Alphei *et al.* 1996; Bonkowski *et al.* 2000b). Introduced bacteria also interact with rhizosphere microfauna. Inoculation of pea (*Pisum sativum*) seeds with strains of the bacterium *Pseudomonas fluorescens* increased the abundance of nematodes and protozoa

in the rhizosphere; non-inoculated germinating pea seedlings exerted a nemati-
cidal effect that was thought to be metabolized and inactivated by the
introduced bacteria (Brimecombe *et al.* 1999). Conversely, inoculation of
wheat (*Triticum aestivum*) with the same bacteria increased rhizosphere populations of nematodes but not protozoa, showing that the outcome of the
plant–microfauna interaction depends on plant characteristics such as root
exudation patterns (Brimecombe *et al.* 1999).

More importantly, the effects of rhizobacteria on root architecture seem to
be controlled to a great extent by protozoan grazing (Bonkowski and Brandt
2002). Plants develop an extensive and more highly branched root system in
the presence of protozoa (Jentschke *et al.* 1995), than when grown in the
absence of protozoa, corresponding to hormonal effects on root growth by
beneficial rhizobacteria (Rolfe *et al.* 1997). Thus, in addition to the stimulation
of gross nutrient flows, protozoa promote a loosely mutualistic interaction
between plant roots and rhizobacteria (Bonkowski and Brandt 2002). Protozoan grazing has been found to promote auxin-producing rhizobacteria,
which stimulates the growth of the root system, allows more nutrients to
be absorbed, and will also increase exudation rates thereby further stimu-
lating bacterial–protozoan interactions, as shown in Figure 3.1 (Bonkowski
and Brandt 2002). These observations have been substantially supported by
the experimental study of protozoan effects on *Arabidopsis thaliana* plants

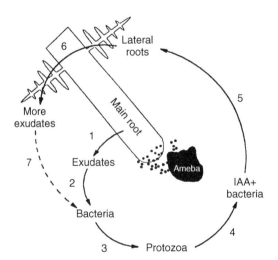

FIGURE 3.1 A conceptual model illustrating microfaunal-induced hormonal effects on root
growth, from Bonkowski and Brandt (2002). Root exudation (1) stimulates growth of a diverse
bacterial community (2) and subsequently of bacterial-feeders such as protozoa (3). Selective feeding by protozoa favors plant hormone–producing bacteria (4) and hormonal release induces lateral
root growth (5), the release of more exudates (6), subsequent bacterial growth (7), and so on.

transformed by the cytokinin-inducible ARR5-promoter-GUS construct. As expected, root elongation and root branching nearly doubled in plants grown in the presence of a naked ameba (*Acanthamoeba castellanii*), compared to control plants grown solely in soil inoculated with a filtered microbial inoculum. Simultaneously, GUS-reporter gene activity strongly increased in treatments with protozoa. The dramatic change in root architecture of *Arabidopsis* suggests a strong auxin effect, which presumably had to be down-regulated in the root by the auxin-antagonist cytokinin. These findings have been summarized in a conceptual model of microfaunal-induced hormonal effects on root growth (Figure 3.1). Effects of microfauna on plant root systems should be integrated with other drivers of roots as detailed in Chapter 10.

Recently the role of other signal molecules, apart from hormones, in microbe–root communication has been established. Phillips *et al.* (1999) found that the bacterium *Sinorhizobium meliloti* produces a signal molecule that enhances root respiration and triggers a compensatory increase in whole-plant net carbon assimilation in alfalfa (*Medicago sativa*). They identified the signal as lumichrome, a common breakdown product of riboflavin. In addition, a large proportion of the bacteria colonizing the roots of plants are capable of producing N-acyl homoserine lactone (AHL) signals to coordinate their behavior in local rhizosphere populations. Specific interactions of bacteria with plant hosts, like nodulation (Wisniewski-Dyé and Downie 2002) or the successful infection of plants by deleterious bacteria, seem to depend on such AHL-mediated "quorum-sensing" regulation. Recently, Mathesius *et al.* (2003) demonstrated that auxin responses and investment in defence by the legume *Medicago truncatula* were directly affected by AHLs from both free-living beneficial and deleterious bacteria. Additionally, Joseph and Phillips (2003) showed that homoserine lactone, the breakdown product of AHL, leads to a strong increase of water transpiration in bean (*Phaseolus vulgaris*) and speculated that the microorganisms would benefit from enhanced transpiration when soil moisture carries mineral nutrients toward the root. These examples demonstrate the presence of several, indirect, plant–microorganism interactions that could potentially be significantly affected by the action of rhizosphere microfauna.

3.4 RHIZOSPHERE MICROFAUNA – INTERACTIONS WITH MYCORRHIZAL AND OTHER SYMBIONTS

The outcome of the symbiosis between mycorrhizal fungi and the plant is normally regarded as positive for the plant, as it is supplied with nutrients from the mycorrhiza (Smith and Read 1997). But different arbuscular-mycorrhizal (AM) fungi affect the growth of individual plants differently (Jakobsen 1992), and at high mycorrhizal infection the AM can be harmful to the plant (Gange and Ayres 1999). The effect of AM on plant parasites aboveground can be beneficial or detrimental to the plant, probably dependent on the availability of phosphorus in the soil (West 1995). Actually, it is possible that the problems often seen with establishment of AM in agricultural crops (Iver Jakobsen, pers. comm.) are to a large extent caused by the dual effect of AM fungi that switch between symbionts and parasites (Gange and Ayres 1999). We will deal here only with interactions between mycorrhiza and microfauna; the wider issues of mycorrhizae are covered in depth in Chapter 6.

Root infection by symbiotic rhizosphere organisms often affects populations of rhizosphere protozoa, especially if the plant is also stressed by environmental factors including elevated concentrations of atmospheric CO_2 (Rønn et al. 2002) or herbivory (Wamberg et al. 2003; see Table 3.2). A likely explanation is that the mycorrhizal fungus can directly access photo-assimilate C from the roots, thereby reducing the flow of C into the rhizosphere as exudates. Differences between mycorrhizal and non-mycorrhizal plants are likely to be exaggerated under conditions increasing C-allocation belowground. Bonkowski et al. (2001b), in an experiment with ecto-mycorrhizae and protozoa in soil, saw a reduction of bacteria and protozoa in the presence of mycorrhizae, and

TABLE 3.2   Combined Effects of Foliar Herbivory (By the Weevil *Sitona lineatus*) and Mycorrhizal Root Infection (By *Glomus intraradices*) on Belowground Carbon Allocation in Pea (*Pisum sativum*) Plant Pre- and Post-Flowering. Means Followed by Different Letters are Significantly Different ($P < 0.05$), Data from Wamberg et al. (2003)

		Mycorrhizae Present		Mycorrhizae Absent	
		Herbivory Present	Herbivory Absent	Herbivory Present	Herbivory Absent
Pre-flowering	Plant dry wt. (g)	1.1	1.2	1.2	1.2
	Relative herbivory	2.7	–	2.0	–
	Mycorrhizal roots (%)	58[a]	28[b]	–	–
	Soil + root respiration ($\mu g\,C\,g^{-1}\,d^{-1}$)	14[a]	18[a]	40[b]	16[a]
	Protozoa (000's g^{-1})	18[a]	21[a]	53[b]	25[a]
Post-flowering	Plant dry wt. (g)	1.2	1.2	1.0	1.2
	Relative herbivory	0.5[a]	–	2.3[b]	–
	Mycorrhizal roots (%)	42[a]	68[b]	–	–
	Soil + root respiration ($\mu g\,C\,g^{-1}\,d^{-1}$)	23	26	29	18
	Protozoa (000's g^{-1})	34	20	38	39

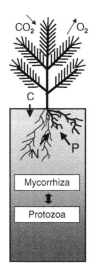

Interactions of Protozoa and Mycorrhiza

Root architecture

	Root number	Root diameter	Root tips	Root length
Mycorrhiza	−40%		−35%	−40%
Protozoa		−25%	+50%	+60%

Trade-off in C allocation

	Fungal hyphae	Bacteria	Amebae
Mycorrhiza		−38%	−34%
Protozoa	−18%		

Labels on diagram: CO_2 O_2 C N P Mycorrhiza Protozoa

FIGURE 3.2   Diagramatic representation of the trade-off in C-allocation between rhizosphere microfauna and mycorrhizal symbionts in *Picea abies*. Data from Bonkowski *et al.* (2001b).

a reduction of fungal mycelium in the presence of protozoa, suggesting a trade-off in C-allocation (Figure 3.2).

Mycorrhizal fungi are known to stimulate leaf-sucking insects, probably because of an increased nutrient content in the sap (Gange *et al.* 1999), but these fungi also reduce the activity of leaf-chewing insects, possibly due to an increased content of structural compounds in the leaves (Gange and West 1994). Foliar herbivory is reported to inhibit mycorrhizal fungi in most cases (Gehring and Whitham 1994). There is a significant interaction with the growth phase of the plant, however, since leaf-chewing insects stimulate mycorrhiza and free-living rhizosphere microorganisms in the early nutrient acquisition phase of the plant, but not during flowering when herbivory did not affect belowground organisms (Wamberg *et al.* 2003; Table 3.2). Grazing by soil fauna is known to affect mycorrhizal fungi, although most work concerns microarthropods (Setälä 1995) and microfaunal effects on mycorrhizal fungi have received less attention. Fungal-feeding nematodes (Yeates *et al.* 1993) and protozoa (Hekman *et al.* 1992) are common in soils so an interaction between microfauna and fungi, including mycorrhiza, is probable. Hánēl and Šimek (1993) observed a significant positive correlation between seasonal changes in plant-feeding nematodes and nitrogen-fixing nodules on red clover (*Trifolium pretense*) roots, and also between bacterial-feeding nematodes and root nitrogenase activity. That soil microfauna reproduced more effectively on some microbial species as food than others is well documented (e.g. Grewel 1991), but the observation that bacterial-feeding nematodes isolated from the potato (*Solanum tuberosum*) rhizosphere reproduced better on a *Comamonas* bacterium than eight other bacterial strains isolated from the potato rhizosphere, and that *Comamonas* is a growth-promoting rhizobacterium, opens the possibility of a specific relationship such that the nematode is favored by and moves the bacterium around the rhizosphere; the bacterium in turn enhances plant growth which stimulates growth of both the bacterium and the nematode (Kimpinski and Sturtz 1996). Interactions between rhizosphere microfauna and plant symbionts other than mycorrhizae are probably more widespread than published studies suggest.

3.5 PLANT RESPONSE TO ABOVE-
AND BELOWGROUND HERBIVORY

The interactions between rhizosphere organisms and foliar herbivores must be mediated by plant responses. Foliar-feeding insects have a variable effect on above- and belowground plant biomass (Wardle 2002), but usually induce an increased carbon flow to the plant roots (Bardgett *et al.* 1998) and a higher root respiration (Holland *et al.* 1996). Defoliation can result in an increased number of nematodes in the rhizosphere (Mikola *et al.* 2001), also indicating increased allocation of plant carbon in the soil. The allocation of plant-C to different microbial interactions varies with the growth stage of the plant and the presence of both foliar herbivores and symbiotic mycorrhiza (Bonkowski *et al.* 2001a; Wamberg *et al.* 2003). The example shown in Table 3.2 shows that in the vegetative, pre-flowering stage, when plants are actively investing C belowground, herbivore-induced increases in C-transfer are used mainly for the production of mycorrhizal fungal biomass (seen by the increases in soil respiration and rhizosphere protozoa if no mycorrhizae are present, and the increase in mycorrhizal infection with foliar-herbivory). In the post-flowering stage, when C-transfer to developing seeds is important, herbivory does increase C-transfer belowground.

Root-feeding insects with different specificity toward crops and weeds (House *et al.* 1984) can have a beneficial or detrimental effect on the growth of a single plant (Gange and Brown 1989; Wardle 2002). In a plant community these insects can increase the N transport from clover to grasses (Hatch and Murray 1994), as detailed for root-feeding nematodes above, and even increase shoot growth for both plant species (Bardgett *et al.* 1999). Root feeding by other invertebrates has been shown to stimulate plant defense chemicals in the leaves, and so reduce populations of foliar-feeding insects

(Bezemer *et al.* 2003). It is not known whether root feeding by nematodes similarly impacts upon foliar plant defenses.

Host-specific root pathogenic nematodes can strongly influence when one plant out competes another and so influence plant succession via rhizosphere effects (Van der Putten *et al.* 2001; Wardle 2002). The presence of root herbivorous nematodes to a large extent depends on plant nutrient status (Verschoor *et al.* 2001) and an increase in number of endoparasitic over ectoparasitic nematodes has been observed in barley grown at low N and P as opposed to fully fertilized barley (Vestergard 2004). Different types of plant parasites may therefore indicate the nutrient status of the host plants.

3.6 CONCLUSIONS AND DIRECTIONS FOR FUTURE RESEARCH

The interactions between plants and microfauna in the rhizosphere are clearly not simply limited to the mineralizing activities of the fauna, nor are they unidirectional with the fauna impactingly solely on the plant. Rather, there are a complex series of interactions between plants, symbiotic flora, fauna, and soil nutrient status with the microfauna affecting, and being affected by, both the shoot and the root portions of the plant. These interactions are also evident at the level of the individual plant as well as the plant community. It is well known that the plants respond to differences in soil nutrient content. Thus, root growth is stimulated in portions of soil with elevated nutrients (Hodge *et al.* 1999) and the root system can also benefit from nutrient pulses of a few hours duration (Campbell and Grime 1989). It is an open question to what extent there has been an evolutionary benefit for the plant of being able to direct its carbohydrates toward the different consumers in direct response to the needs of the plant. The soil, fauna, flora, root, shoot, herbivores, and predators in many ways act like a single, connected organism. Rhizosphere microfauna provide a useful focus in the study of the complex interactions. Future significant advances in understanding and management will come from a holistic approach to the "rhizo-organism."

ACKNOWLEDGEMENTS

The Scottish Crop Research Institute receives grant-in-aid from the Scottish Executive Environment and Rural Affairs Department.

REFERENCES

Alphei, J., M. Bonkowski, and S. Scheu. 1996. Protozoa, Nematoda and Lumbricidae in the rhizosphere of *Hordelymus europaeus* (Poaceae): Faunal interactions, response of microorganisms and effects on plant growth. *Oecologia* 106:111–126.

Bardgett, R.D., D.A. Wardle, and G.W. Yeates. 1998. Linking above ground and below ground interactions: How plant responses to foliar herbivory influence soil organisms. *Soil Biol. Biochem.* 30:1867–1878.

Bardgett, R.D., C.S. Denton, and R. Cook. 1999. Below ground herbivory promotes soil nutrient transfer and root growth in grassland. *Ecol. Lett.* 2:357–360.

Bezemer, T.M., R. Wagenaar, N.M. van Dam, and F.L. Wackers. 2003. Interactions between above- and belowground insect herbivores as mediated by the plant defense system. *Oikos* 101:555–562.

Boenigk, J., and H. Arndt. 2002. Bacterivory by heterotrophic flagellates: Community structure and feeding strategies. *Antonie van Leeuwenhoek* 81:465–480.

Bonkowski, M., and F. Brandt. 2002. Do soil protozoa enhance plant growth by hormonal effects? *Soil Biol. Biochem.* 34:1709–1713.

Bonkowski, M., W. Cheng, B.S. Griffiths, J. Alphei, and S. Scheu. 2000a. Microbial-faunal interactions in the rhizosphere and effects on plant growth. *Eur. J. Soil Biol.* 36:135–147.

Bonkowski, M., B.S. Griffiths, and C. Scrimgeour. 2000b. Substrate heterogeneity and microfauna in soil organic "hotspots" as determinants of nitrogen capture and growth of rye-grass. *Appl. Soil Ecol.* 14:37–53.

Bonkowski, M., I.E. Geoghegan, A.N.E. Birch, and B.S. Griffiths. 2001a. Effects of soil decomposer invertebrates (protozoa and earthworms) on an above-ground phytophagous insect (cereal aphid), mediated through changes in the host plant. *Oikos* 95:441–450.

Bonkowski, M., G. Jentschke, and S. Scheu. 2001b. Contrasting effects of microbes in the rhizosphere: interactions of mycorrhiza (*Paxillus involutus* (Batsch) Fr.), naked amoebae (Protozoa) and Norway Spruce seedlings (*Picea abies* Karst.). *Appl. Soil Ecol.* 18:193–204.

Brimecombe, M.J., F.A.A.M. De Leij, and J.M. Lynch. 1999. Effect of introduced *Pseudomonas fluorescens* strains on soil nematode and protozoan populations in the rhizosphere of wheat and pea. *Microb. Ecol.* 38:387–397.

Campbell, B.D., and J.P. Grime. 1989. A comparative study of plant responsiveness to the duration of episodes of mineral nutrient enrichment. *New Phytol.* 112:261–268.

Cipollini, D., C.B. Purrington, and J. Bergelson. 2003. Costs of induced responses in plants. *Basic Appl. Ecol.* 4:79–85.

Clarholm, M. 1985. Interactions of bacteria, protozoa and plants leading to mineralization of soil nitrogen. *Soil Biol. Biochem.* 17:181–187.

Coûteaux, M.M., G. Faurie, L. Palka, and C. Steinberg. 1988. Le relation prédateur-proie (protozoaires-bactéries) dans les soils: Rôle dans la régulation des populations et conséquences sur les cycles du carbonne et de l'azote. *Rev. Ecol. Biol. Sol* 25:1–31.

Curl, E.A., and J.D. Harper. 1990. Fauna-microflora interactions. In: Lynch, J.M. (ed.) *The Rhizosphere*. John Wiley & Sons, Chichester, pp. 369–388.

Gange, A.C., and R.L. Ayres. 1999. On the relation between arbuscular mycorrhizal colonization and plant "benefit." *Oikos* 87:615–621.

Gange, A.C., and V.K. Brown. 1989. Insect herbivory affects size variability in plant populations. *Oikos* 56:351–356.

Gange, A.C., and H.M. West. 1994. Interactions between arbuscular mycorrhizal fungi and foliar feeding insects in *Plantago lanceolata* L. *New Phytol.* 128:79–87.

Gange, A.C., E. Bower, and V.K. Brown. 1999. Positive effects of an arbuscular mycorrhizal fungus on aphid life history traits. *Oecologia* 120:123–131.

Gehring, C.A., and T.G. Whitham. 1994. Interactions between aboveground herbivores and the mycorrhizal mutualists of plants. *Trends Ecol. Evolut.* 9:251–255.

Grewel, P.S. 1991. The influence of bacteria and temperature on the reproduction of *Caenorhabdidtis elegans* (Nematoda: Rhabditidae) infesting mushrooms (*Agaricus bisporus*). *Nematologica* 37:72–82.

Griffiths, B.S. 1990. A comparison of microbial-feeding nematodes and protozoa in the rhizosphere of different plants. *Biol. Fertil. Soils* 9:83–88.

Griffiths, B.S. 1994. Soil nutrient flow. In: Darbyshire, J.F. (ed.) *Soil Protozoa*. CAB International, Wallingford, pp. 65–91.

Griffiths, B.S., and D. Robinson. 1992. Root-induced nitrogen mineralization, a nitrogen balance model. *Pl. Soil* 139:253–263.

Griffiths, B.S., I.M. Young, and B. Boag. 1991. Nematodes associated with the rhizosphere of barley (*hordeum vulgare*). *Pedobiologia* 35:265–272.

Griffiths, B.S., M. Bonkowski, G. Dobson, and S. Caul. 1999. Changes in soil microbial community structure in the presence of microbial-feeding nematodes and protozoa. *Pedobiologia* 43:297–304.

Hánĕl, L., and M. Šimek. 1993. Soil nematodes and nitrogenase activity (symbiotic N_2-fixation) in red-clover (*Trifolium pratense* L.). *Eur. J. Soil Biol.* 29:109–116.

Hatch, D.J., and P.J. Murray. 1994. Transfer of nitrogen from damaged roots of white clover (*Trifolium repens* L.) to closely associated roots of intact perennial ryegrass (*Lolium perenne* L.). *Pl. Soil* 166:181–185.

Hekman, W.E., P.J.H.F. van den Boogert, and K.B. Zwart. 1992. The physiology and ecology of a novel, obligate mycophagous flagellate. *FEMS Microbiol. Ecol.* 86:255–265.

Hodge, A., J. Stewart, D. Robinson, B.S. Griffiths, and A.H. Fitter. 1999. Plant, soil fauna and microbial responses to N rich organic patches of contrasting temporal availability. *Soil Biol. Biochem.* 31:1517–1530.

Holland, M.A. 1997. Occam's razor applied to hormonology: Are cytokinins produced by plants? *Plant Physiol.* 115:865–868.

Holland, J.N., W. Cheng, and D.A. Crossley. Jr. 1996. Herbivore induced changes in plant carbon allocation: Assessment of below ground C fluxes using carbon 14. *Oecologia* 107:87–94.

House, G.J., B.R. Stinner, D.A. Crossley, Jr., and E.P. Odum. 1984. Nitrogen cycling in conventional and no tillage agroecosystems: Analysis of pathways and processes. *J. Appl. Ecol.* 21:991–1012.

Ingham, R.E., J.A. Trofymow, E.R. Ingham, and D.C. Coleman. 1985. Interactions of bacteria, fungi, and their nematode grazers: Effects on nutrient cycling and plant growth. *Ecol. Monogr.* 55:119–140.

Ingham, E.R., J.A. Trofymow, R.N. Ames, H.W. Hunt, C.R. Morley, J.C. Moore, and D.C. Coleman. 1986. Trophic interactions and nitrogen cycling in a semi arid grassland soil 1. Seasonal dynamics of the natural populations, their interactions and effects on nitrogen cycling J. Appl Ecol. 23:597–614.

Jakobsen, I. 1992. Phosphorus transport by external hyphae of vesicular arbuscular mycorrhizas. In: Read, D.J., Lewis, D.H., Fitter, A.H., and Alexander, I.J. (eds) *Mycorrhizas in Ecosystems*. CAB International, Wallingford, UK, pp. 48–54.

Jentschke, G., M. Bonkowski, D.L. Godbold, and S. Scheu. 1995. Soil protozoa and forest tree growth: Non-nutritional effects and interaction with mycorrhizae. *Biol. Fertil. Soils* 20:263–269.

Joseph, C.M., and D.A. Phillips. 2003. Metabolites from soil bacteria affect plant water relations. *Pl. Physiol. Biochem.* 41:189–192.

Kaye, J.P., and S.C. Hart. 1997. Competition for nitrogen between plants and soil microorganisms. *Trends Ecol. Evolut.* 12:139–143.

Kimpinski, J., and A.V. Sturtz. 1996. Population growth of a Rhabditid nematode on plant growth promoting bacteria from potato tubers and rhizosphere soil. *J. Nematol.* (suppl.) 28S:682–686.

Lambrecht, M., Y. Okon, A. Vande Broek, and J. Vanderleyden. 2000. Indole-3-acetic acid: A reciprocal signalling molecule in bacteria–plant interactions. *Trends Microbiol.* 8:298–300.

Lynch, J.M., and J.M. Whipps. 1990. Substrate flow in the rhizosphere. *Pl. Soil* 129:1–10.

McKenzie Bird, D., and H. Koltai. 2000. Plant parasitic nematodes: Habitats, hormones, and horizontally-acquired genes. *J. Pl. Growth Reg.* 19:183–194.

Marilley, L., and M. Aragno. 1999. Phylogenetic diversity of bacterial communities differing in degree of proximity of *Lolium perenne* and *Trifolium repens* roots. *Appl. Soil Ecol.* 13:127–136.

Mathesius, U., S. Mulders, M.S. Gao, M. Teplitski, G. Caetano-Anolles, B.G. Rolfe, and W.D Bauer. 2003. Extensive and specific responses of a eukaryote to bacterial quorum-sensing signals. *Proc. Natl. Acad. Sci.* 100:1444–1449.

Mengel, K. 1996. Turnover of organic nitrogen in soils and its availability to crops. *Pl. Soil* 181:83–93.

Mikola, J., G.W. Yeates, G.M. Barker, D.A. Wardle, and K.I. Bonner. 2001. Effects of defoliation intensity on soil food web properties in an experimental grassland community. *Oikos* 92:333–343.

Patten, C.L., and B.R. Glick. 1996. Bacterial biosynthesis of indole-3-acetic acid. *Can. J. Microbiol.* 42:207–220.

Phillips, D.A., C.M. Joseph, G.P. Yang, E. Martínez-Romero, J.R. Sanborn, and H. Volpin. 1999. Identification of lumichrome as a *Sinorhizobium* enhancer of alfalfa root respiration and shoot growth. *Proc. Natl. Acad. Sci.* 96:12275–12280.

Phillips, D.A., and D.R. Strong. 2003. Rhizosphere control points: Molecules to food webs. *Ecology* 84:815.

Posch, T., K. Šimek, J. Vrba, S. Pernthaler, J. Nedoma, B. Sattler, B. Sonntag, and R. Psenner. 1999. Predator-induced changes of bacterial size-structure and productivity studied on an experimental microbial community. *Aquat. Microb. Ecol.* 18:235–246.

Rolfe, B.G., M.A. Djordjevic, J.J. Weinman, U. Mathesius, C. Pittock, E. Gartner, K.M. Ride, Z.M. Dong, M. McCully, and J. McIver. 1997. Root morphogenesis in legumes and cereals and the effect of bacterial inoculation on root development. *Pl. Soil* 194:131–144.

Rønn, R., M. Gavito, J. Larsen, I. Jakobsen, H. Frederiksen, and S. Christensen. 2002. Response of free living soil protozoa and microorganisms to elevated atmospheric CO_2 and presence of mycorrhiza. *Soil Biol. Biochem.* 34:923–932.

Rovira, A.D. 1991. Rhizosphere research- 85 years of progress and frustration. In: Kleister, D.L., and Cregan, P.B. (eds) *The Rhizosphere and Plant Growth.* Kluwer Academic Publishers, The Netherlands, pp. 3–13.

Ryle, G.J.A., C.E. Powell, and A.J. Gordon. 1979. Respiratory costs of nitrogen-fixation in soybean, cowpea, and white clover.1. Nitrogen-fixation and the respiration of the nodulated root. *J. Exp. Bot.* 30:135–144.

Semenov, A.M., A.H.C. van Bruggen, and V.V. Zelenev. 1999. Moving waves of bacterial populations and total organic carbon along roots of wheat. *Microb. Ecol.* 37:116–128.

Setälä, H. 1995. Growth of birch and pine-seedlings in relation to grazing by soil fauna on ectomycorrhizal fungi. *Ecology* 76:1844–1851.

Shishido, M., H.B. Massicotte, and C.P. Chanway. 1996. Effect of plant growth promoting *Bacillus* strains on pine and spruce seedling growth and mycorrhizal infection. *Ann. Bot.* 77:433–441.

Smith, S.E., and D.J. Read. 1997. *Mycorrhizal Symbiosis.* Academic Press, London.

Van der Putten, W.H., L.E.M. Vet, J.A. Harvey, and Wäckers, F.L. 2001. Linking above and belowground multitrophic interactions of plants, herbivores, pathogens, and their antagonists. *Trends Ecol. Evolut.* 16:547–554.

Verhagen, F.J.M., H.J. Laanbroek, and J.W. Woldendorp. 1995. Competition for ammonium between plant-roots and nitrifying and heterotrophic bacteria and the effects of protozoan grazing. *Pl. Soil* 170:241–250.

Verschoor, B.C., R.G.M. de Goede, F.W. de Vries, and L. Brussaard. 2001. Changes in the composition of the plant-feeding nematode community in grasslands after cessation of fertiliser application. *Appl. Soil Ecol.* 17:1–17.

Vestergard, M. 2004. Nematode assemblages in the rhizosphere of spring barley (*Hordeum vulgare* L.) depended on fertilisation and plant growth phase. *Pedobiologia* 48:257–265.

Wamberg, C., S. Christensen, and I. Jakobsen. 2003. Interactions between foliar-feeding insects, mycorrhizal fungi, and rhizosphere protozoa on pea plants. *Pedobiologia* 47:281–287.

Wang, J.G., and L.R. Bakken. 1997. Competition for nitrogen during decomposition of plant residues in soil: Effect of spatial placement of N-rich and N-poor plant residues. *Soil Biol. Biochem.* 29:153–162.

Wardle, D.A. 2002. Communities and Ecosystems, Linking aboveground and belowground components. *Monographs in Population Ecology* 34, Princeton University Press.

West, H.M. 1995. Soil phosphate status modifies response of mycorrhizal and non mycorrhizal *Senecio vulgaris* L. to infection by the rust, *Puccinia lagenophorae* Cooke. *New Phytol.* 129:107–116.

Wisniewski-Dye, F., and J.A. Downie. 2002. Quorum-sensing in *Rhizobium*. *Antonie van Leeuwenhoek* 81:397–407.

Yeates, G.W., T. Bongers, R.G.M. de Goede, D.W. Freckman, and S.S. Georgieva. 1993. Feeding-habits in soil nematode families and genera – An outline for soil ecologists. *J. Nematol.* 25:315–333.

Yeates, G.W., S. Saggar, C.S. Denton, and C.F. Mercer. 1998. Impact of clover cyst nematode (*Heterodera trifolii*) infection on soil microbial activity in the rhizosphere of white clover (*Trifolium repens*) – A pulse-labelling experiment. *Nematologica* 44:81–90.

Yeates, G.W., S. Saggar, C.B. Hedley, and C.F. Mercer. 1999. Increase in C-14-carbon translocation to the soil microbial biomass when five species of plant-parasitic nematodes infect roots of white clover. *Nematology* 1:295–300.

Zwart, K.B., P.J. Kuikman, and J.A. van Veen. 1994. Rhizosphere protozoa: Their significance in nutrient dynamics. In: Darbyshire, J.F. (ed.) *Soil Protozoa*, CAB International, Wallingford, pp. 93–122.

CHAPTER 4

Mycorrhizas: Symbiotic Mediators of Rhizosphere and Ecosystem Processes

Nancy C. Johnson and Catherine A. Gehring

4.1 INTRODUCTION

Roots of most terrestrial plants form symbiotic associations with fungi. These ubiquitous symbioses, called mycorrhizas, function as conduits for the flow of energy and matter between plants and soils. The term "mycorrhizosphere" was coined to describe the unique properties of the rhizosphere surrounding and influenced by mycorrhizas (Linderman 1988). Figure 4.1 illustrates pine seedlings with and without mycorrhizas to highlight some of these properties. Mycorrhizal fungi frequently stimulate plants to reduce root biomass while simultaneously expanding nutrient uptake capacity by extending far beyond root surfaces and proliferating in soil pores that are too small for root hairs to enter. Mycelial networks of mycorrhizal fungi often connect plant root systems over broad areas. These fungi frequently comprise the largest portion of soil microbial biomass (Olsson *et al.* 1999; Högberg and Högberg 2002). Thus, mycorrhizal symbioses physically and chemically structure the rhizosphere, and they impact communities and ecosystems.

Excellent recent reviews of mycorrhizal biology (Smith and Read 1997), physiology (Kapulnik and Douds 2000), evolution (Brundrett 2002), and ecology (van der Heijden and Sanders 2002; Read and Perez-Moreno 2003) are available. The purpose of this chapter is to examine the roles of mycorrhizas in the structure and functioning of communities and ecosystems, and to explore their responses to anthropogenic environmental changes.

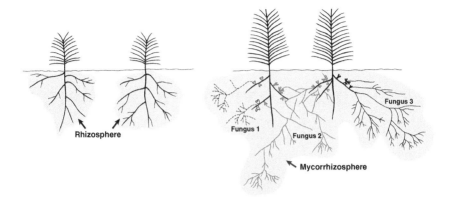

FIGURE 4.1 Drawing of rhizosphere versus mycorrhizosphere. The rhizosphere (left) and mycorrhizosphere (right) of ectomycorrhizal (EM) pine seedlings differ dramatically from one another in plant and soil attributes. Mycorrhizal plants typically have larger shoots and smaller root systems compared to non-mycorrhizal plants. The extent of the soil explored by fungi vastly exceeds that of the root system, though the average magnitude of this effect is underemphasized in this illustration. More than one species of EM fungi frequently colonizes a host plant imparting its own unique properties on the portion of the root system it colonizes. In this figure, hypothetical fungal species 1 is shown in dashed lines, 2 in gray, and 3 in solid black. Fungi 1, 2, and 3 may differ in their drought tolerance, ability to utilize organic nutrient sources, extent of soil exploration, ability to compete with saprotrophs, energetic cost to the plant or in a variety of other ways. Fungal species 2 is shared by two conspecific plants and may allow exchange of resources between adjacent plants. In addition to these differences, EM fungi alter rhizosphere chemistry through unique fungal compounds such as chitin, and ergosterol, and through the production of diverse carbohydrates, enzymes, organic acids, and secondary metabolites. The combined physical, chemical, and biotic changes associated with the mycorrhizosphere influence the fitness of individual pine seedlings and also have ecosystem-scale consequences.

4.2   CONVERGENT EVOLUTION OF MYCORRHIZAS

Throughout their evolution, plant roots have repeatedly formed symbioses with fungi. With remarkably few exceptions, plant roots have evolved to accommodate, utilize and control mycorrhizal fungi. Both molecular and fossil evidence indicate that the earliest land plants were mycorrhizal (Redecker *et al.* 2000). These bryophytic plants did not possess true roots but rather stem-like rhizomes that were colonized with fungi that appear similar to modern-day arbuscular mycorrhizal (AM) fungi (Stubblefield *et al.* 1987). Pirozynski and Malloch (1975) suggest that plants could not have colonized land without fungal partners capable of acquiring nutrients from the undeveloped soils that existed during the Silurian and Devonian. Once terrestrial plants became established and soil organic matter accrued, more mycorrhizal partnerships evolved as plant and fungal taxa radiated into the newly forming terrestrial niches rich in organic matter. These disparate symbioses have been grouped into six general types of mycorrhizas: arbuscular (also called vesicular–arbuscular), ecto, ericoid, arbutoid, monotropoid, and orchid (Table 4.1; Smith and Read 1997).

Mycorrhizas are highly variable in structure, yet they have evolved two common features: an elaborate interface between plant root and fungal cells, and extraradical hyphae that extend into the soil. This chapter will focus primarily on arbuscular, ecto-, and to a limited extent, ericoid mycorrhizas. However, a brief examination of the similarities and differences of all six types of mycorrhizas reveals points of evolutionary convergence and divergence of mycorrhizal symbioses.

ARBUSCULAR MYCORRHIZAS

Arbuscular mycorrhizas are widespread and abundant. They are formed by bryophytes, pteridophytes, gymnosperms, and angiosperms, and are ubiquitous in most temperate and tropical ecosystems including agricultural systems. The fungal partners in AM associations are remarkably abundant, accounting from 5 to 50 percent of the microbial biomass in agricultural soils (Olsson *et al.* 1999). These fungi are members of the Glomeromycota, a monophyletic phylum containing 150–160 described species (see Table 4.1). Arbuscular mycorrhizas are sometimes called "endomycorrhizas" because the fungal partner forms intraradical structures (i.e., inside plant roots). In AM associations, the interface between plant and fungal tissues that facilitates exchange of materials between plant and fungal symbionts takes the form of arbuscules or coils. Arbuscules and coils are modified fungal hyphae that provide a large surface area for resource exchange. Several genera of AM fungi also form intraradical vesicles that function as fungal storage organs. The extraradical hyphae of AM fungi lack regular cross walls allowing materials, including nuclei, to flow relatively freely within the mycelium. These hyphae can be very abundant; one gram of grassland soil may contain as much as 100 m of AM hyphae (Miller *et al.* 1995). The taxonomy of AM fungi is based upon the morphology of large (10–600 μm diameter) asexual spores produced in the soil or within roots.

ECTOMYCORRHIZAS

Ectomycorrhizas occur in certain families of woody gymnosperms (e.g., Pinaceae) and angiosperms (e.g., Dipterocarpaceae, Betulaceae) and are extremely important in many temperate and boreal forests. The fungal partners in ectomycorrhizal (EM) associations account for an estimated 30 percent

TABLE 4.1 Characteristics of Six Distinct Groups of Mycorrhizas

Mycorrhiza	Fungal partners	Plant partners	Resources exchanged from plant/from fungus	Ecosystems where mycorrhiza predominates
Arbuscular	Glomeromycota: Glomales	Bryophytes, Pteridophytes, Gymnosperms, Angiosperms	CHO/mineral forms of P, N, Zn, Cu, etc.	Agroecosystems, grasslands, deserts, temperate deciduous forests, tropical rainforests
Ectomycorrhiza	Basidiomycota, Ascomycota, and Zygomycota	Gymnosperms, Angiosperms	CHO/organic & mineral forms of N, P, Zn, Cu, etc.	Boreal forests, evergreen and deciduous temperate forests
Ericoid	Ascomycota: Leotiales	Ericales: Ericaceae, Epacridaceae, Empetraceae	CHO/organic & mineral forms of N, P, Zn, Cu, etc.	All ecosystems where the Ericales occur, particularly abundant in tundra, heathlands, boreal forests
Arbutoid	Basidiomycota	Ericales: *Arbutus*, *Arctostaphylos*, Pyrolaceae, Bryophytes	CHO/organic & mineral forms of N, P, Zn, Cu, etc.	Chaparral and other ecosystems where the Arbutoideae occur
Monotropoid	Basidiomycota	Ericales: Monotropoideae	CHO (from photosynthetic plant)/CHO (to heterotrophic plant), minerals	Evergreen and deciduous temperate forests of the northern hemisphere
Orchid	Basidiomycota	Orchidaceae	? from plant/CHO and minerals from fungus	All ecosystems where orchids occur, particularly abundant in tropical systems

Sources of information: Smith and Read 1997; Brundrett 2002.

of the microbial biomass in forest soils (Högberg and Högberg 2002). These fungi are a diverse assemblage of at least 6000 species of basidiomycetes, ascomycetes, and zygomycetes (Table 4.1; Smith and Read 1997). The oldest fossils providing clear evidence of EM associations date back 50 million years (LePage *et al.* 1997), yet the association is hypothesized to have evolved 130 million years ago (Smith and Read 1997).

Structurally, ectomycorrhizas are characterized by the presence of a fungal mantle that envelops host roots and a Hartig net that surrounds root epidermal and/or cortical cells and provides a large surface area for resource exchange. Hormonal interactions between plant and fungus lead to dramatically altered root architecture including the suppression of root hairs. The external component of EM associations consists of hyphae with cross walls that partition cellular components. These hyphae sometimes coalesce into macroscopic structures called rhizomorphs that attach the mycelium to sporocarps or can be morphologically similar to xylem and serve in water uptake (Smith and Read 1997). The external mycelium of EM fungi may be more extensive than that of AM fungi with as much as 200 m of hyphae per gram of dry soil (Read and Boyd 1986). Ectomycorrhizal fungi also are frequently classified using the morphology of colonized roots and their sporocarps, such as the familiar mushrooms and truffles.

MYCORRHIZAS IN THE ERICALES

The plant order Ericales contains a natural group of closely related families with worldwide distribution. Plants in this order form three distinctive forms of mycorrhizas: ericoid, arbutoid, and monotropoid (Table 4.1). Ericoid mycorrhizas involve partnerships between ascomycetes and members of the Ericaceae, Epacridaceae, and Empetraceae families. In the ericoid mycorrhizas, the epidermal cells of small-diameter roots lack root hairs and instead are frequently filled with fungal hyphae. Arbutoid mycorrhizas form between basidiomycetes and members of the Pyrolaceae and some genera of Ericaceae, most notably *Arbutus* and *Arctostaphylos*. Structurally, arbutoid mycorrhizas are similar to ectomycorrhizas as they possess a thick fungal mantle and a Hartig net, yet they are characterized by the formation of dense hyphal complexes within root epidermal cells. Monotropoid mycorrhizas are partnerships between certain non-photosynthetic members of the Monotropaceae and basidiomycetes. In these associations, the fungus transfers carbohydrates from a photosynthetic plant to its achlorophyllous (myco-heterotrophic) host plant. In addition to a fungal mantle and Hartig net, these mycorrhizas are characterized by a "peg" of fungal hyphae that proliferates within the epidermis of the root (Smith and Read 1997).

ORCHID MYCORRHIZAS

Members of the Orchidaceae form a unique type of mycorrhizas with some basidiomycetes (Table 4.1). Orchids differ from other plants because they pass through a prolonged seedling (protocorm) stage during which they are unable to photosynthesize and are dependent upon a fungal partner to supply exogenous carbohydrate (Smith and Read 1997). Adult plants of most species of orchids are green and photosynthetic, but an estimated 200 species remain achlorophyllous throughout their life. These orchids are considered to be "myco-heterotrophic" because they acquire fixed carbon heterotrophically through their mycorrhizal fungal partner (Leake 1994). Orchid mycorrhizas are morphologically distinct as well, consisting of intracellular hyphae that form a complex interface between plant and fungal symbionts termed a peloton. Smith and Read (1997) and Leake (1994) question whether or not these associations should be even considered mycorrhizas because there is no demonstrated benefit of the association to the fungus.

4.3   MYCORRHIZAS AS NUTRITIONAL MUTUALISMS

Except for orchid and monotropoid associations, mycorrhizas involve plant exchange of photosynthates in return for fungal exchange of mineral nutrients. The convergence of so many unrelated forms of mycorrhizas is a testament for the mutual benefits of these trading partnerships. To understand the dynamics of resource exchange in mycorrhizas, we must examine the mechanisms by which resources are acquired by both partners. Mycorrhizal fungi improve nutrient uptake for plants, in part, by exploring the soil more efficiently than plant roots. Mycorrhizal fungal hyphae occupy large volumes of soil, extending far beyond the nutrient depletion zone that develops around roots. Simard *et al.* (2002) estimated that, on average, the external hyphae of EM fungi produce a 60-fold increase in surface area. The small diameter of fungal hyphae allows them to extract nutrients from soil pore spaces too small for plant roots to exploit (van Breemen *et al.* 2000). Recent studies on phosphate and ammonium uptake also reveal that mycorrhizal fungi improve uptake kinetics through reductions in K_m and increases in V_{max} (van Tichelen and Colpaert 2000).

Most mycorrhizal fungi depend heavily on plant photosynthate to meet their energy requirements; AM fungi are obligate biotrophs while EM and ericoid fungi are biotrophs with some saprotrophic abilities. The carbon cost of mycorrhizas is difficult to accurately estimate, but field and laboratory studies suggest that plants allocate 10–20 percent of net primary production to their fungal associates (Smith and Read 1997). Root colonization by mycorrhizal fungi often increases rates of host plant photosynthesis. This effect has been attributed to mycorrhizal enhancement of plant nutritional status in some systems (Black *et al.* 2000) and a greater assimilate sink in other systems (Dosskey *et al.* 1990).

Mycorrhizal fungi are a significant carbon sink for their host plants, and if nutrient uptake benefits do not outweigh these carbon costs, then both plant and fungal growth can be depressed (Peng *et al.* 1993; Colpaert *et al.* 1996). Mycorrhizal biomass has been shown to both increase and decrease with increasing availability of soil nitrogen (Wallenda and Kottke 1998; Johnson *et al.* 2003a). Treseder and Allen (2002) proposed a conceptual model to account for this apparent contradiction (Figure 4.2a). The model is based on three premises:

1. Both plants and mycorrhizal fungi have minimum N and P requirements and plants have a higher total requirement for these nutrients than fungi.
2. Biomass of mycorrhizal fungi is limited by the availability of plant carbon allocated belowground.

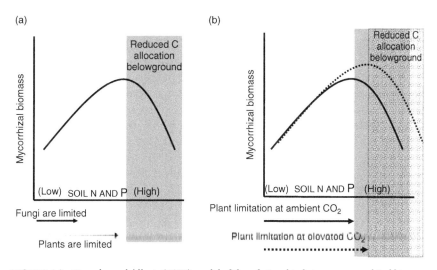

FIGURE 4.2   Treseder and Allen's (2002) model of the relationship between mycorrhizal biomass and soil nutrients. At very low levels of soil N and P, mycorrhizal biomass will increase with nutrient enrichment until plant hosts are no longer limited by these resources. When N and P levels are sufficient for plants, then mycorrhizal biomass is expected to decrease with additional nutrients because fungi will become carbon limited as plants reduce carbon allocation below-ground – shown in the shaded area (a). Mycorrhizal biomass is predicted to increase with elevated atmospheric carbon dioxide because plant demands for N and P will rise as carbon assimilation rates increase, and mycorrhizal fungi will be less carbon limited. Thus, at elevated carbon dioxide (dotted line) the mycorrhizal biomass response curve will be higher and shifted to the right compared to ambient carbon dioxide (solid line) (b).

3. Plants allocate less photosynthate belowground when they are not limited by nitrogen and phosphorus; thus, mycorrhizal growth decreases
when availability of these nutrients is high.

At very low soil nitrogen and phosphorus availability, both plants and mycorrhizal fungi are nutrient limited, so enrichment of these resources will increase
mycorrhizal growth. At very high nitrogen and phosphorus availability, nei-
ther plants nor fungi are limited by these elements; consequently mycorrhizal
biomass is reduced as plants allocate relatively less photosynthate belowground and more aboveground to shoots (shaded area in Figure 4.2a). This
model is useful because it provides a simple heuristic framework for understanding how the relative availability of below- (minerals) and aboveground
(photosynthate) resources control mycorrhizal biomass. Considering the interplay between nitrogen and phosphorus availability may further enhance the
predictive value of this model. Because mycorrhizal fungi generally acquire
phosphorus more readily than their host plants, we predict that the mutualistic value of mycorrhizal associations to plants will be highest at high soil N:P
ratios and diminish as N:P ratios decrease.

Two lines of evidence suggest that mycorrhizal plants have evolved mechanisms to actively balance photosynthate costs with mineral nutrient benefits.
First, environmental factors that reduce photosynthetic rates, such as low
light intensity, lead to reductions in mycorrhizal development (e.g., Gehring
2003). Secondly, plant allocation to root structures is sensitive to mycorrhizal
benefits. This is observed at both a gross taxonomic level and within ecotypes
of the same plant species. Plant taxa with coarse root systems (low surface
area) are generally more dependent upon mycorrhizas than those with fibrous
root systems (high surface area). This suggests that for highly mycotrophic
plant taxa, it is more adaptive to provide a fungal partner with photosynthates
than to maintain fibrous root systems (Newsham *et al.* 1995). Also, it appears
that mycotrophic plants have evolved a certain degree of plasticity in their
allocation to roots in response to their mycorrhizal status. Mycorrhizal plants
often have reduced root:shoot ratios compared to non-mycorrhizal plants of
the same species grown under identical conditions (Mosse 1973; Colpaert
et al. 1996; Figure 4.1).

There is evidence that local ecotypes of plants and mycorrhizal fungi
co-adapt to each other and to their local soil environment (Figure 4.3a).
A comparison of *Andropogon gerardii* ecotypes from phosphorus-rich and
phosphorus-poor prairies show that each ecotype grew best in the soil of its
origin. Furthermore, the *A. gerardii* ecotype from the phosphorus-poor soil
was three times more responsive to mycorrhizal colonization and had a significantly coarser root system than the ecotype from the phosphorus-rich soil
(Schultz *et al.* 2001). These results suggest that the genetic composition of

(a) (b)

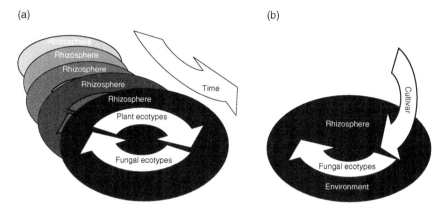

FIGURE 4.3 Ecotypes of co-occurring plant and mycorrhizal fungi are expected to evolve in response to each other and their local rhizosphere environment (a). Agriculture, horticulture, and plantation forestry uncouple evolutionary feedbacks between plant and mycorrhizal fungal ecotypes (b).

plant populations evolve so that mycorrhizal costs are minimized and benefits are maximized within the local soil fertility conditions.

4.4 COMMUNITY INTERACTIONS

Mycorrhizal interactions influence the species composition, diversity, and stability of biotic communities. Assessing mycorrhizal roles in communities is challenging because the ubiquity and abundance of these associations makes it difficult to remove them from intact communities so that their function can be accurately measured. Nevertheless, experiments using microcosms (e.g., van der Heijden *et al.* 1998), selective fungicides (e.g., Hartnett and Wilson 1999), and theoretical and empirical studies (e.g., Bever *et al.* 1997) indicate that mycorrhizal feedbacks are a significant force in structuring plant communities.

Variation among mycorrhizal associations in resource acquisition is important to rhizosphere dynamics. Species of mycorrhizal fungi vary in the degree to which they explore the soil with extraradical hyphae (Erland and Taylor 2002; Hart and Reader 2002). Species of EM fungi have been shown to vary more than threefold in their nutrient uptake rates (Colpaert *et al.* 1999) suggesting large differences in their effects on both host plant performance and rhizosphere nutrient cycling. Intraspecific variation can also be substantial as different strains of the same mycorrhizal fungal species can vary more than different species in aspects of nutrient uptake (Cairney 1999; Graham and Abbott 2000).

MYCORRHIZAL FEEDBACKS ON PLANT COMMUNITY STRUCTURE

A community model developed by Bever (Bever *et al.* 1997) assumes that the population growth rates of plants and mycorrhizal fungi are mutually interdependent and identifies the potential for two very different community dynamics. Symmetrical delivery of benefits between plants and fungi will generate a positive feedback, and asymmetrical delivery of benefits will generate a negative feedback. Positive feedback strengthens the mutualism between individual pairs of plants and fungi, yet decreases community diversity; while negative feedback weakens the mutualism between individual plant–fungus pairs and maintains community diversity. Recent experiments indicate that both of these mechanisms occur within natural communities, and that variation in the balance of mycorrhizal costs and benefits may be extremely important in structuring plant communities. Klironomos (2002) found that mutualistic isolates of AM fungi were selected for within the rhizospheres of individual plants grown in pots during two 10-week cycles. In contrast, when 10 plant species were randomly paired with 10 AM fungal isolates from the same grassland, the function of the partnerships varied from strongly mutualistic to strongly parasitic (Klironomos 2003). These studies indicate that co-adaptation of plant–fungus pairs occurs at the centimeter scale within individual plant rhizospheres, not at the hectare scale within grassland swards. Furthermore, this work provides solid experimental support for the hypothesis that ecotypes of plants and mycorrhizal fungi co-adapt to one another and to their local soil environment (Figure 4.3a), and this process may be an important determinant of community structure.

Extraradical hyphae from individual clones of mycorrhizal fungi frequently link the root systems of neighboring plants of the same as well as different species (Figure 4.1). In this way, most mycorrhizal plants are interconnected by a common mycorrhizal network (CMN) at some point in their life. Isotope labeling studies show that carbon and mineral nutrients can be transferred among neighboring plants within this CMN; however, the magnitude and rate of this transfer appears to vary greatly between AM and EM systems as well as among plant and fungal taxa (Simard *et al.* 2002). Although it is well established that interplant transfer of carbon and nutrients occurs, there is debate over whether the amount of material transferred is large enough to affect plant physiology and ecology, and whether the materials leave the fungal tissues in the roots and reenter the shoots of the receiver plant (Simard *et al.* 2002). In this regard, EM and AM associations appear to differ. In a field study using dual $^{13}C/^{14}C$ labeling, Simard *et al.* (1997) showed significant bidirectional shoot-to-shoot carbon transfer between adjacent *Pseudotsuga* and *Betula* seedlings colonized by a common EM fungus. In contrast, Fitter *et al.* (1998) found that although AM fungi transferred a significant amount of carbon between the root systems of the grass *Cynodon* and the forb *Plantago*, this carbon was never released into the receiving plant's shoots. Thus, Fitter *et al.* (1998) suggest that interplant movement of carbon via common AM mycelia is less likely to impact plant fitness than AM fungal fitness. There is a great need for field-based research to test the claims that CMNs influence seedling survival, assist species recovery following disturbance, influence plant diversity by altering the competitive balance of plant species, reduce nutrient loss from ecosystems, and increase productivity and stability of ecosystems. Simard *et al.* (2002) review studies that both support and contradict these claims. Future research will help resolve the role of CMNs in community structure and ecosystem processes.

BACTERIA–MYCORRHIZA INTERACTIONS

Effects of mycorrhizal fungi on the quantity and quality of root exudates may generate a cascade of effects on populations and communities of rhizosphere bacteria (see Chapter 3; Linderman 1988). Recent data suggest that different combinations of plant–fungal pairs generate different effects on bacterial communities. For example, Söderberg *et al.* (2002) observed that the outcome of AM colonization by *Glomus intraradices* on bacterial communities varied among plant species. While some studies suggest that mycorrhizal fungi and soil bacteria may compete for carbon in the rhizosphere, other studies indicate that plant growth promotion by mycorrhizal fungi may counteract this effect and actually stimulate rhizosphere bacterial activity (Söderberg *et al.* 2002).

Some bacteria, such as a number of fluorescent pseudomonads, are known to function as mycorrhization helper bacteria (MHB) because of their ability to consistently enhance mycorrhizal development (reviewed by Garbaye 1994). Rates of ectomycorrhiza formation on *Eucalyptus diversicolor* in plant nurseries were increased up to 300 percent by MHBs (Dunstan *et al.* 1998), leading to significant increases in seedling biomass. The mechanisms for these effects are poorly understood, but the release of volatile compounds by MHBs has been shown to stimulate fungal growth (Garbaye 1994). Available data suggest that these interactions are extremely complex. For example, MHBs isolated from the *Pseudotsuga menziesi–Laccaria laccata* symbiosis were fungus-specific, but not plant-specific. These bacteria promoted EM establishment of the fungus *L. laccata* but inhibited the formation of ectomycorrhizas by other species of fungi (Garbaye 1994).

FUNGAL–MYCORRHIZA INTERACTIONS

Mycorrhizal fungi co-inhabit the rhizosphere with many saprotrophic and pathogenic fungi. Interactions with saprotrophic fungi may be more likely for the EM and ericoid fungi that have significant abilities to degrade organic matter. Based on a trenching study, Gadgil and Gadgil (1971) suggested that EM fungi might reduce decomposition rates by competing with saprotrophic fungi for nutrients. Microcosm studies in which saprotrophic and mycorrhizal mycelial systems are allowed to interact provide evidence for significant retardation of the growth of saprotrophic fungi by EM fungi and vice versa (e.g., Leake *et al.* 2002). Ectomycorrhizal fungi may outcompete saprotrophic fungi for rhizosphere territory (Lindahl *et al.* 2001), release organic acids into the rhizosphere that inhibit saprotrophs (Rasanayagam and Jeffries 1992), or indirectly reduce litter decomposition by saprotrophs by extracting water from the soil (Koide and Wu 2003).

Fungal pathogens and mycorrhizal fungi also interact with one another. Many studies demonstrate significant protection from pathogens by mycorrhizal fungi. In a meta-analysis of studies of interactions among AM fungi and fungal pathogens, Borowicz (2001) showed that plants generally grow better when they are mycorrhizal and this is especially true when plants are challenged by pathogens. Newsham *et al.* (1995) suggest that pathogen suppression may be more important than other benefits of AM symbioses in natural ecosystems. The mechanisms of pathogen suppression are highly varied and include improved nutrition of the host plant, changes in the chemical composition of plant tissues, and changes in rhizosphere bacterial communities (Linderman 2000).

ANIMAL–MYCORRHIZA INTERACTIONS

Interactions among soil animals and mycorrhizal fungi may potentially alter the function of the symbiosis (see Chapter 3). Many studies of relationships among soil fauna and mycorrhizal fungi focus on the impacts of fungus-feeding animals, such as collembola, on mycorrhizal function. Hiol *et al.* (1994) found that grazing by the collembolan *Proisotoma minuta* significantly reduced EM colonization and also that the collembolan had distinct preferences for the hyphae of certain species of EM fungi. Several studies suggest that hyphal grazers prefer to feed on saprotrophic or parasitic fungi over mycorrhizal fungi and thus may have limited impact on the symbiosis (e.g., Hiol *et al.* 1994). Even when hyphal grazing results in substantial reductions in mycorrhizal fungal hyphae, the mutualism may not be negatively affected. For example, Setälä (1995) found that although soil fauna reduced EM fungal biomass by as much as 50 percent, the growth of both birch and pine seedlings was 1.5–1.7-fold greater in the microcosms with soil fauna compared to those without.

Non-intuitive outcomes of interactions between soil fauna and mycorrhizal fungi reveal how little is known about the natural history of most soil organisms. Klironomos and Hart (2001) showed that the EM fungus *L. laccata* actively kills and consumes the collembolan *Folsomia candida* and the host plant benefited from this unconventional foraging behavior of its EM fungus. Stable isotope labeling showed that up to 25 percent of the nitrogen found in *Pinus stroba* seedlings colonized by *L. laccata* was of collembolan origin. Mycorrhizal fungi also affect the growth of plant pathogenic nematodes and generally reduce the detrimental effects that these animals have on plant growth. Borowicz (2001) found that AM fungi significantly impact nematode growth and that this effect varies with the type of nematode. Sedentary nematodes are negatively affected by AM fungi while migratory species actually grow better in their presence.

The studies that we have reviewed here were selected to demonstrate that complex interactions among communities of mycorrhizal fungi and other soil organisms can mediate rhizosphere processes. Yet these studies represent only a small sample of the myriad of interactions among mycorrhizal fungi and other rhizosphere organisms. For example, we did not discuss the role of animals such as earthworms in dispersing fungal propagules or the indirect effects of root herbivores on mycorrhizal fungi (Gange and Brown 2002). Likewise, we have limited our discussion to belowground interactions despite evidence that mycorrhizal symbioses are influenced by aboveground organisms such as herbivores (e.g., Gehring and Whitham 2002). The study of interactions between mycorrhizal fungi and other organisms is important not only to our understanding of rhizosphere processes, but also to the broader study of relationships such as competition because of the altered dynamics possible because of the carbon subsidy provided to these fungi by plants.

4.5 ECOSYSTEM INTERACTIONS AND BIOGEOGRAPHY

SOIL STRUCTURE

One of the most important functions of mycorrhizas is their role in physically structuring soils. This is significant because soil structure mediates fertility, water content, root penetration, and erosion potential of soils. Mycorrhizas generate stable soil aggregates. Miller and Jastrow (2000) suggested that mycorrhizas physically and chemically bind soil particles into stable macroaggregates like "sticky string bags" of mycorrhizal hyphae and associated roots. Furthermore, in many but not all soil types, mycorrhizal hyphae are involved with the hierarchical arrangement of macro- and microaggregates within the soil matrix. The contributions of mycorrhizas to soil structure vary with soil type and also with plant and fungal phenotype. For example, AM fungi generally play a greater role in aggregate formation in sandy soil than in clayey soil (Miller and Jastrow 2000). Although both saprotrophic and mycorrhizal fungi facilitate the formation of soil aggregates, mycorrhizal fungi stabilize soil much more effectively than saprotrophic fungi. Miller and Jastrow (2000) propose three reasons for this: (1) mycorrhizal fungi have direct access to plant photosynthate, and consequently they are less carbon limited than saprotrophic fungi; (2) hyphae of mycorrhizal fungi are often more persistent than hyphae of saprotrophic fungi; and (3) hyphae of ericoid, EM, and AM fungi exude sticky glycoproteinaceous slimes that facilitate the binding of particles within the mycorrhizal "string bags." Glomalin is a glycoprotein that has been linked with stability of soil aggregates (Miller and Jastrow 2000). Wright *et al.* (1996) used a monoclonal antibody to detect glomalin on the surfaces of actively growing hyphae from representatives of five different genera of AM fungi and also, to a lesser extent, on the hyphae of some non-AM fungi. Glomalin is a putative homolog of heat shock protein 60 (Gadkar and Rillig 2006). Further studies are needed to thoroughly characterize this important soil-binding agent.

DECOMPOSITION AND NUTRIENT CYCLING

The total effect of roots and associated mycorrhizal fungi on soil structure is related to their turnover and decomposition rates. Fungal hyphae contain high amounts of chitin, a compound known to be resistant to decomposition. Although it is not yet well characterized, glomalin also appears to be remarkably recalcitrant. Rillig *et al.* (2001) used radiocarbon dating to estimate a residence time of 6–42 years! Langley and Hungate (2003) concluded that mycorrhizas can be important controllers of root decomposition rates, and also that the tissues and exudates formed in EM associations are more recalcitrant than those formed in AM associations. The species of mycorrhizal fungi is an important consideration. Wallander and colleagues found that five morphotypes of EM fungi on *Pinus sylvestris* varied more than twofold in chitin concentration (Wallander *et al.* 1999). Future studies are clearly needed to assess the relative importance of plant and fungal control of mycorrhizal influences on soil development.

Mycorrhizal fungi also influence nutrient cycling by accessing nutrients from inorganic and organic sources that are generally unavailable to plants directly. Ectomycorrhizal fungi secrete organic acids into the rhizosphere, and thus increase the weathering of soil minerals (van Breemen *et al.* 2000; see Chapter 8). Although AM fungi appear to have little ability to degrade complex organic molecules, EM and ericoid mycorrhizas can derive nitrogen, phosphorus, and sulfur through the enzymatic breakdown of complex organic compounds (Read and Perez-Moreno 2003).

Traditional models of nutrient cycling consider soil microbes as carbon-limited saprotroph that provide plants with inorganic nutrients in the soil. This view misses key aspects of nutrient cycling in ecosystems dominated by ericoid and EM mycorrhizas, and several authors have argued for a more mycocentric view of nutrient cycling. Lindahl *et al.* (2002) suggest that mineralization and the inorganic nutrients that result from it are relatively unimportant to nutrient cycling in many coniferous forests. Instead, organic sources of nutrients predominate in these ecosystems, and plant access to these nutrients depends upon the outcome of interactions between decomposers and EM fungi. In this view, EM and ericoid mycorrhizal fungi acquire some nutrients from the soil directly, and also capture organic nutrients from plant litter and, perhaps more importantly, from other soil biota including mycorrhizal and saprotrophic fungi, bacteria, protozoa, and soil microfauna (Klironomos and Hart 2001). Furthermore, EM fungi can sequester large quantities of nitrogen in their external mycelia. These nitrogen stores can be mobilized and utilized by host plants when nitrogen demands are high, for example, during bud break in the spring.

GLOBAL PATTERNS OF MYCORRHIZAS

Aerts (2002) suggested that variation in access to different nutrient pools among types of mycorrhizal fungi leads to positive feedbacks between the dominant plants in an ecosystem and litter decomposition. These feedbacks, combined with abiotic environmental constraints, can shape ecosystems at broad spatial scales including the global distribution of biomes. Distinctive types of mycorrhizas correspond with the climatic and edaphic conditions that characterize the major terrestrial biomes (Read 1991). Ericoid mycorrhizas are most common at the highest latitudes and altitudes, EM associations dominate boreal forests and temperate coniferous forests, and AM associations are abundant in low-latitude biomes (Figure 4.4). This pattern relates to soil fertility constraints imposed by pedogenic processes (Read 1991). Cold and wet conditions at high latitudes and altitudes slow decomposition. Organic matter accumulates in these regions and nitrogen and phosphorus are largely contained within organic compounds (e.g., chitin, proteins, and amino acids) and soil humus. Consequently, it is extremely adaptive for EM and ericoid mycorrhizas within tundra and boreal biomes to be capable of assimilating nitrogen and phosphorus from organic sources. As latitude decreases, mean annual temperature increases along with decomposition rates and soil pH. These changes influence resource availability and also enzymatic activity. Read (1991) considered the pH optima of the enzymes utilized by EM fungi and

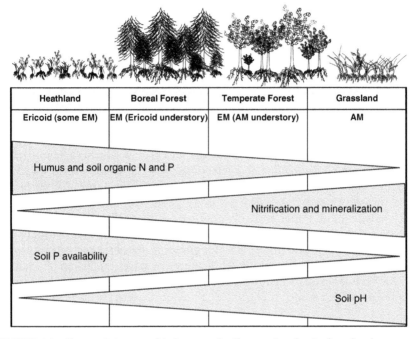

Heathland	Boreal Forest	Temperate Forest	Grassland
Ericoid (some EM)	EM (Ericoid understory)	EM (AM understory)	AM

Humus and soil organic N and P

Nitrification and mineralization

Soil P availability

Soil pH

FIGURE 4.4   Characteristic mycorrhizal types and soil properties of major boreal and temperate biomes. Mycorrhizal fungi with advanced saprotrophic capabilities predominate in high latitude and altitude biomes where decomposition and mineralization processes are inhibited. Modified from Read and Perez-Moreno 2003.

suggested that the progressive replacement of EM by AM associations as soil pH increases above 5 reflects the loss of the selective advantage conferred by EM mobilization of organic nutrients. The efficiency of AM fungi in acquiring inorganic phosphorus corresponds with the general pattern for increasing phosphorus limitation with decreasing latitude (Read 1991; Read and Perez-Moreno 2003; Figure 4.4).

4.6   MYCORRHIZAL FUNCTION IN A CHANGING WORLD

It is well recognized that humans are changing global environments at an unprecedented rate. These changes are known to impact global climate and biota; however, the ramifications for communities and ecosystems are not yet known (IPCC 2001). Understanding mycorrhizal responses to anthropogenic environmental changes can help predict the trajectories of future communities and ecosystems in a changing world.

RESOURCE ENRICHMENT

Human activities have more than doubled annual inputs of nitrogen into the biosphere (Vitousek *et al.* 1997) and increased atmospheric carbon dioxide concentrations by 31 percent since 1750 (IPCC 2001). This enrichment clearly impacts root systems as discussed in Chapter 7. Mycorrhizal responses to nitrogen and carbon dioxide enrichment have also been reviewed (e.g., Wallenda and Kottke 1998; Rillig *et al.* 2002); here we will explore the mechanisms of these responses.

Enrichment of soil nitrogen and atmospheric carbon dioxide can be expected to alter the balance of trade between mycorrhizal fungi and their plant hosts. When phosphorus, water, and other soil resources are not in limiting supply, then nitrogen enrichment is predicted to reduce mycorrhizal biomass as host plants reduce carbon allocation to roots and fungal partners (Figure 4.2a). Using the same conceptual model, we can expect that carbon dioxide enrichment should increase mycorrhizal biomass because plant demands for nitrogen and phosphorus will increase concurrently with carbon assimilation rates, and plants will allocate more photosynthate belowground to roots and mycorrhizal fungi to help satisfy the increased demand for nitrogen and phosphorus (Figure 4.2b). A serious limitation of these conceptual models is that they are inherently phytocentric because they assume that plant allocation of photosynthate is the sole controller of mycorrhizal development. An alternative more mycocentric view has been proposed by Wallander (1995) in which the fungus, rather than the plant, adjusts allocation patterns in response to resource availability.

Many studies support the prediction that nitrogen enrichment should decrease mycorrhizal biomass (Figure 4.2a); however, there is considerable variability in this response. The biomass of reproductive structures and extraradical mycelia of mycorrhizal fungi are more consistently reduced by nitrogen enrichment than is intraradical colonization. Results of long-term field studies show that nitrogen enrichment dramatically affects sporocarp production by EM fungi (Wallenda and Kottke 1998) and spore production by AM fungi (Egerton-Warburton and Allen 2000). Long-term nitrogen enrichment of a spruce forest in Sweden reduced the growth of EM mycelia by 50 percent (Nilsson and Wallander 2003). Similarly, long-term nitrogen enrichment consistently reduced AM hyphal lengths in North American grasslands with sufficient soil phosphorus (Johnson *et al.* 2003a). Formation of EM root tips and intraradical AM colonization have been shown to decrease, increase, or stay the same in response to nitrogen enrichment (Wallenda and Kottke 1998; Johnson *et al.* 2003a). This variability suggests that plant phenotypes, fungal phenotypes, and edaphic conditions mediate mycorrhizal responses to nitrogen enrichment and cautions against extrapolating the results from one system to other systems.

Fungal taxa clearly differ in their responses to resource enrichment. Some species of mycorrhizal fungi decline with nitrogen enrichment while others proliferate (e.g., Egerton-Warburton and Allen 2000; Lilleskov *et al.* 2001). Wallenda and Kottke (1998) reviewed the literature and concluded that EM fungal species with a narrow host range (particularly conifer specialists) are more adversely affected than species with a broad range of host plants. When soil phosphorus is not limiting, members of the AM fungal family Gigasporaceae are often dramatically reduced by nitrogen enrichment (Egerton-Warburton and Allen 2000; Johnson *et al.* 2003a). In contrast, when soil phosphorus is in limiting supply, nitrogen enrichment increases AM fungal biomass and especially populations of Gigasporaceae (Johnson *et al.* 2003a). This suggests that nitrogen enrichment of phosphorus-deficient soils exacerbates phosphorus limitation and increases the net benefits of mycorrhizas. Gigasporaceae populations seem particularly sensitive to plant responses to changing soil N:P ratios.

The conceptual model presented in Figure 4.2b predicts that elevated atmospheric carbon dioxide should increase mycorrhizal biomass as carbon becomes relatively less limiting and soil nutrients become relatively more limiting to plant growth. Some studies support this model, while others do not. For example, elevated carbon dioxide has been shown to increase extraradical hyphal lengths of both AM and EM fungi (Rillig *et al.* 1999; Treseder and Allen 2000). In contrast, other studies have found that when plant size is factored out, there are no effects of carbon dioxide on EM (Walker *et al.* 1997) or AM root colonization (Staddon *et al.* 1999). As with mycorrhizal responses to nitrogen enrichment, it is likely that these apparent contradictions arise from differences among the experimental systems, and, in particular, differences in the taxa of interacting plants and fungi.

The species composition of EM fungi in a spruce forest in northern Sweden responded dramatically to carbon dioxide enrichment (Fransson *et al.* 2001). Similarly, AM fungal communities have been shown to change in response to carbon dioxide enrichment (Wolf *et al.* 2003). Klironomos and colleagues (2005) demonstrated that abruptly increasing atmospheric carbon dioxide levels from 350 to 550 ppm in a single step changed the species composition and functioning of AM fungal communities. However, these researchers also showed that responses did not occur when carbon dioxide level was increased gradually over 21 generations. This study provides sobering evidence that experiments applying a single-step increase in carbon dioxide may be overestimating some community responses to carbon dioxide enrichment.

CLIMATE CHANGE

Average global temperature has risen since the early 1980s, and the Intergovernmental Panel on Climate Change predicts land surface temperature will increase 3.1° C by 2085 (IPCC 2001). Likely changes in average precipitation vary throughout the world, yet the probability of extreme precipitation events is expected to rise. For example, drought frequency and severity is expected to increase over most mid-continental land interiors in the coming decades (IPCC 2001). Effects of altered temperature and precipitation regimes on mycorrhizas are difficult to predict for two reasons. First, these changes will influence both above- and belowground environments, and thereby have the potential to both directly and indirectly impact mycorrhizas (Rillig *et al.* 2002). Second, temperature and precipitation changes are frequently linked to one another so that realistic scenarios of global change must include both factors along with the importance of altered temporal patterns such as changes in growing season length or the timing of precipitation.

Rillig *et al.* (2002) concluded that temperature increases in the ranges predicted by climate models may promote mycorrhizal fungi directly through temperature-dependent increases in fungal metabolism and indirectly through increases in plant growth and nutrient mineralization in the soil. These changes were expected to be most dramatic in regions near the poles where potential increases in nutrient mineralization could favor AM fungi in areas formerly dominated by ericoid or EM associations. When combined with the changes predicted for nitrogen enrichment, these patterns further suggest a world of increasing AM dominance as human impacts increase.

More variable precipitation regimes are also likely to affect mycorrhizal fungi. Root colonization by mycorrhizal fungi can respond significantly to soil moisture content (e.g., Swaty *et al.* 1998) and the relationship may be nonlinear and variable depending upon the taxa of plants and fungi involved. Mycorrhizal fungi have been shown to affect plant–water relations and drought tolerance in a number of AM and EM hosts (e.g., Boyle and Hellenbrand 1991; Auge 2001). Species and even isolates of mycorrhizal fungi vary in their ability to tolerate dry conditions and to assist their hosts in doing so (Stahl and Smith 1984). Simulated drought stress of beech (*Fagus sylvaticus*) resulted in significant shifts in EM community composition and increases in the production of sugar alcohols that are thought to play a role in compensation for drought stress (Shi *et al.* 2002). Querejeta *et al.* (2003) demonstrated that host plants may sustain mycorrhizal fungal hyphae during times of drought through hydraulic lift of moisture from deeper soil layers occupied by roots but not fungal hyphae. The maintenance of a functional mycorrhizal mycelium will not only provide benefits to the host plant and fungus but also to a variety of other rhizosphere organisms.

The drought tolerance of certain taxa of plants and fungi will be exceeded as climates continue to change. Plants are expected to be less tolerant of water stress than fungi because fungi can access smaller soil pore spaces and survive remarkably low water potentials. Some fungi are among the most xerotolerant organisms known (Kendrick 2000), though the drought tolerance of mycorrhizal fungi has not been broadly tested. Because of their dependence on host plant carbon, when drought is extreme, mycorrhizal fungi may share the same fate as their host plants regardless of their individual drought tolerance. For example, recent droughts in southwestern North America have resulted in substantial mortality of pinyon pines (*Pinus edulis*). Surviving pinyons in high-mortality areas supported a less abundant, less diverse, and compositionally different EM community than neighboring pinyons growing in low-mortality sites on similar soils (Swaty *et al.* 2004). Furthermore, survival rates of pinyon seedlings were 50 percent lower in high-mortality sites that were depauperate in EM fungi compared to sites with high populations of EM fungi. Because pinyons are the only hosts for EM fungi in many of these habitats, their loss from the system may also mean loss of EM fungi from large tracts of woodland. Interestingly, AM fungi predominate in most water-limited desert environments. Here again, predicted environmental changes appear to favor AM fungi over EM fungi.

Increasingly, studies of relationships between mycorrhizal fungi and global change are focusing on interactions among multiple factors rather than single factors (Rillig *et al.* 2002). This is an important advance because few environments will experience only one change, and interactions among multiple factors may generate complex outcomes. For example, increased carbon dioxide availability may reduce the consequences of drought and also change the balance between costs and benefits in mycorrhizal symbioses. This was shown in a recent mesocosm experiment (Johnson *et al.* 2003b). Elevated carbon dioxide increased the species richness of the plant community when AM fungi were present but not when they were absent. The survival of slow-growing mycotrophic forbs was increased when carbon dioxide was enriched suggesting that the treatment ameliorated the carbon cost of the symbiosis (Johnson *et al.* 2003b). Anthropogenic loss of biodiversity and the introduction of exotic species will likely further complicate these responses.

BIODIVERSITY CHANGES

There is growing concern about the consequences of the recent loss of global biodiversity caused by human activities. Conventional high-input agriculture is an extreme case of anthropogenic reduction of biodiversity. As human population increases, vast areas of diverse natural communities have been

replaced with crops, orchards, and plantation forests consisting of genetically uniform cultivars. These changes clearly impact mycorrhizas and other
rhizosphere organisms (see Chapter 6). When compared to adjacent natural
areas, communities of AM fungi in cultivated fields are generally less diverse
and dominated by a few agriculture-tolerant taxa (Helgason et al. 1998). If
agriculture-tolerant taxa of mycorrhizal fungi are effective mutualists, then
the changes in fungal communities that accompany cultivation may benefit
crop production. Unfortunately, there is no theoretical or empirical evidence
to support this scenario (Ryan and Graham 2002). Rather, there is evidence
supporting the hypothesis that many modern agricultural practices may inadvertently generate less mutualistic communities of AM fungi (Johnson et al.
1992). For example, high levels of mineral fertilizer cause plants to reduce
allocation to roots and mycorrhizas. Because AM fungi are obligate biotrophs,
reduced availability of host carbon is a very strong selection pressure, and
fungal phenotypes that best commandeer plant carbon will have a strong
advantage over less-aggressive phenotypes. This aggressive acquisition of host
carbon is clearly adaptive for fungi living in high-fertility soils; however, it
also predisposes them to be less mutualistic, or even parasitic on their host
plants (Johnson et al. 1992; Kiers et al. 2002).

Communities of plants and rhizosphere organisms are continually adapting
to one another in undisturbed ecosystems (Figure 4.3a). In contrast, the plant
phenotypes that occur in production agriculture are selected by farmers and
not by their ability to maximize the benefits and minimize the costs of their
rhizosphere symbioses (Figure 4.3b). Crop breeding programs may exacerbate
the proliferation of inferior mycorrhizal mutualists by selecting cultivars that
perform best with high fertilizer inputs, and consequently with low levels
of mycorrhizal colonization. Older land races of crops are consistently more
mycotrophic than modern cultivars (Hetrick et al. 1993). Efforts to incorporate
mycorrhizas into more sustainable agricultural programs need to recognize
the mycotrophic properties of the crops as well as the mutualistic properties
of the fungi.

Complementary to our concern over global loss of species diversity is the
growing threat of exotic species that can significantly alter community and
ecosystem processes. Mycorrhizal fungi may indirectly enhance the success of
some exotic plant invaders (e.g., Marler et al. 1999), but our understanding
of the potential effects of mycorrhizal fungi as exotic species is rudimentary
owing to our poor knowledge of the diversity and species composition of
mycorrhizal fungal communities in any ecosystem.

One example of unintended consequences from the introduction of an
exotic pine and associated EM fungi illustrates the potential importance of
fungal species diversity and/or species composition to ecosystem processes
(Chapela et al. 2001). The establishment of pine plantations in many parts

of the world requires the concomitant introduction of compatible EM fungi. These alien plant–fungal combinations are introduced to novel environments where they can inadvertently alter community and ecosystem processes. In this example, a species-poor community of three EM fungi associated with one host tree, Monterey pine (*Pinus radiata*), was introduced into a formerly AM-dominated grassland in Ecuador. It should be stressed that the three introduced fungi comprise only a small component of the approximately 100 EM fungi that occur in native Monterey pine forests. In less than 20 years, this introduced EM-pine system removed up to 30 percent of soil carbon stored in these grasslands (Chapela *et al.* 2001). Introduction of this plant–fungus combination altered rhizosphere carbon pools and thus the carbon cycle in dramatic and unpredicted ways. While we might be tempted to think that, as mutualists, exotic mycorrhizal fungi may not have the dramatic negative impacts of introduced fungal pathogens, some types of mycorrhizal fungi have flexibility in their trophic capabilities, and under the right environmental conditions many of them have the potential to act as parasites.

4.7   DIRECTIONS FOR FUTURE RESEARCH

To date, most studies of mycorrhizal mediation of rhizosphere processes have examined individual plant–fungus pairs or interactions among individual mycorrhizas and other rhizosphere biota or abiotic conditions. Although this scale of inquiry provides precise understanding of specific plant–fungal systems, it cannot provide meaningful information about mycorrhizal function within communities and ecosystems (Read and Perez-Moreno 2003). Also, we still have much to learn regarding the extent of mycorrhizal fungal diversity. Among species of mycorrhizal fungi, there is very little knowledge of functional attributes such as stress tolerance, demand for photosynthate, and nutrient uptake efficiency. It is critical to gain a clearer understanding of functional variation among fungal species to guide conservation and restoration efforts.

In addition to a focus on a small subset of plants and their associated fungi, many mycorrhizal studies apply a phytocentric focus. Incorporation of a mycocentric view is necessary to predict the long-term function of these interactions because mycorrhizal fungi are continually evolving to maximize their own fitness rather than the benefits that they convey to their host plants. Fungi might thus be expected to forage optimally for host plant resources, potentially associating preferentially with individual plants or plant species that are best able to provide fixed carbon. Future research efforts are needed to study the foraging behavior of mycorrhizal fungi and to develop methods to measure fungal fitness. Although it is now possible to define the spatial distribution of some fungal individuals in the field (Hirose *et al.* 2004), few studies have tracked individual fungi through time as environments change. Yet, these kinds of studies are necessary to understanding of the dynamics of mycorrhizal symbioses.

A fungal perspective is also important when considering larger scale impacts of mycorrhizas. For example, work by Gadgil and Gadgil (1971) and Langley and Hungate (2003) showed that the presence of mycorrhizal fungi can alter rates of above- and belowground litter decomposition due to chemical changes in roots and interactions with decomposer fungi. These changes in decomposition rate are likely to influence plants and other members of the community. Measurements of the functioning of the symbiosis should thus go beyond plant growth and fungal biomass if we are to understand these broader consequences.

Study of the impacts of mycorrhizal fungi on rhizosphere processes also would benefit from increased linkage between empirical ecologists and scientists from other disciplines. For example, much of the research on gene expression profiles during the development of the symbiosis is performed under highly controlled conditions that do not mimic the field. Addition of experimental treatments in which carbon and nutrient supply varied could make these studies even more informative. Likewise, ecologists conducting field and greenhouse studies of the impacts of environmental change on mycorrhizal colonization and community composition could benefit from collaboration with geneticists and physiologists to provide mechanistic insights. Increased effort to model the mycorrhizal symbiosis is also important because theoretical approaches can yield new perspectives and generate testable hypotheses.

The analytical challenge of holistic studies of complex interactions among communities of rhizosphere organisms and the environment is daunting. Nevertheless, meeting this challenge is particularly important as we seek to manage mycorrhizas in forestry, agriculture, and in the restoration of highly disturbed areas where these complex interactions may be altered to the detriment of rhizosphere function. In the spirit of G. Evelyn Hutchinson's "evolutionary play in an ecological theater," mycorrhizas are an evolutionary play performed in the rhizosphere theater by a cast of myriads of plant and fungal ecotypes (Figure 4.3a). By cultivating plant ecotypes that have been selected in the absence of endemic mycorrhizal fungi, humans have inadvertently stopped the feedback mechanism in this evolutionary play (Figure 4.3b). Additionally, the ecological theater has been altered by anthropogenic changes to the environment. Understanding long-term mycorrhizal function in communities and ecosystems requires a holistic perspective that considers both the evolutionary play and the ecological theater.

ACKNOWLEDGEMENTS

We appreciate insightful suggestions from Pål-Axel Olsson, Håkan Wallander, Jim Graham, and anonymous reviewers. Financial support was provided by the National Science Foundation DEB-0415563, DEB-0316136, and the Fulbright Commission.

REFERENCES

Aerts, R. 2002. The role of various types of mycorrhizal fungi in nutrient cycling and plant competition. In: van der Heijden, M.G.A., and Sanders, I. (eds) *Mycorrhizal Ecology*. Springer, Berlin, Heidelberg, New York. 117–134 pp.

Auge, R.M. 2001. Water relations, drought and vesicular-arbuscular mycorrhizal symbiosis. *Mycorrhiza* 11:3–42.

Bever, J.K., K.M. Westover, and J. Antonovics. 1997. Incorporating the soil community into plant population dynamics: the utility of the feedback approach. *J Ecol* 85:561–573.

Black, K.G., D.T. Mitchell, and B.A. Osborne. 2000. Effect of mycorrhizal-enhanced leaf phosphate status on carbon partitioning, translocation and photosynthesis in cucumber. *Plant Cell Environ* 23:797–809.

Borowicz, V. 2001. Do arbuscular mycorrhizal fungi alter plant-pathogen relations? *Ecology* 82:3057–3068.

Boyle, C.D., and K.E. Hellenbrand. 1991. Assessment of the effect of mycorrhizal fungi on drought tolerance of conifer seedlings. *Can J Bot* 69:1764–1771.

Brundrett, M.C. 2002. Coevolution of roots and mycorrhizas of land plants. *New Phytol* 154:275–304.

Cairney, J.W.G. 1999. Intraspecific physiological variation: implications for understanding functional diversity in ectomycorrhizal fungi. *Mycorrhiza* 9:125–135.

Chapela, I.H., L.J. Osher, T.R. Horton, and M.R. Henn. 2001. Ectomycorrhizal fungi introduced with exotic pine plantations induce soil carbon depletion. *Soil Biol Biochem* 33:1733–1740.

Colpaert, J.V., A. van Laere, and J.A. van Assche. 1996. Carbon and nitrogen allocation in ectomycorrhizal and non-mycorrhizal *Pinus sylvestris* L. seedlings. *Tree Phys* 16:787–793.

Colpaert, J.V., K.K. van Tichelen, J.A. van Assche, and A. van Laere. 1999. Short-term phosphorus uptake rates in mycorrhizal and non-mycorrhizal roots of intact *Pinus sylvestris* seedlings. *New Phytol* 143:589–597.

Dosskey, M.G., R.G. Linderman, and L. Boersma. 1990. Carbon-sink stimulation of photosynthesis in Douglas fir seedlings by some ectomycorrhizas. *New Phytol* 115:269–274.

Dunstan, W.A., N. Malajczuk, and B. Dell. 1998. Effects of bacteria on mycorrhizal development and growth of container grown *Eucalyptus diversicolor* F. Muell. seedlings. *Plant Soil* 201:241–249.

Egerton-Warburton, L.M., and E.B. Allen. 2000. Shifts in arbuscular mycorrhizal communities along an anthropogenic nitrogen deposition gradient. *Ecol Appl* 10:484–496.

Erland, S., and A.F.S. Taylor. 2002. Diversity of ecto-mycorrhizal fungal communities in relation to the abiotic environment. In: van der Heijden, M.G.A., and Sanders, I. (eds) *Mycorrhizal Ecology*. Springer, Berlin, Heidelberg, New York. 163–200 pp.

Fitter, A.H., J.D. Graves, N.K. Watkins, D. Robinson, and C. Scrimgeour. 1998. Carbon transfer between plants and its control in networks of arbuscular mycorrhizas. *Functional Ecology* 12:406–412.

Fransson, P.M.A., A.F.S. Taylor, and R.D. Finlay. 2001. Elevated atmospheric CO_2 alters root symbiont community structure in forest trees. *New Phytol* 152:431–442.

Gadgil, R.L., and P.D. Gadgil. 1971. Mycorrhiza and litter decomposition. *Nature* 233:133.

Gadkar, V., and M.C. Rillig. 2006. The arbuscular mycorrhizal fungal protein glomalin is a putative homolog of heat shock protein 60. *FEMS Microbiology Letters* 263:93–101.

Gange, A.C., and V.K. Brown. 2002. Actions and interactions of soil invertebrates and arbuscular mycorrhizal fungi in affecting the structure of plant communities. In: van der Heijden, M.G.A., Sanders, I. (eds) *Mycorrhizal Ecology*. Springer, Berlin, Heidelberg, New York. 321–344 pp.

Garbaye, J. 1994. Helper bacteria – a new dimension to the mycorrhizal symbiosis. *New Phytol* 128:197–210.

Gehring, C.A. 2003. Growth responses to arbuscular mycorrhizas by rain forest seedlings vary with light intensity and tree species. *Plant Ecol* 167:127–139.

Gehring, C.A., and T.G. Whitham. 2002. Mycorrhizae-herbivore interactions: population and community consequences. In: van der Heijden, M.G.A., and Sanders, I. (eds) *Mycorrhizal Ecology*. Springer, Berlin, Heidelberg, New York. 295–320 pp.

Graham, J.H., and L.K. Abbott. 2000. Functional diversity of arbuscular mycorrhizal fungi in the wheat rhizosphere. *Plant Soil* 220:179–185.

Hart, M.M., and R.J. Reader. 2002. Taxonomic basis for variation in the colonization strategy of arbuscular mycorrhizal fungi. *New Phytol* 153:335–344.

Hartnett, D.C., and G.W.T. Wilson. 1999. Mycorrhizae influence plant community structure and diversity in tallgrass prairie. *Ecology* 80:1187–1195.

Helgason, T., T.J. Daniell, R. Husband, A.H. Fitter, and J.P.W. Young. 1998. Ploughing up the wood-wide web? *Nature* 394:431.

Hetrick, B.A.D., G.W.T. Wilson, and T.S. Cox. 1993. Mycorrhizal dependence of modern wheat cultivars and ancestors: a synthesis. *Can J Bot* 71:512–518.

Hiol, F.H., R.K. Dixon, and E.A. Curl. 1994. The feeding preference of a mycophagous collembolan varies with the ectomycorrhiza symbiont. *Mycorrhiza* 5:99–103.

Hirose, D., J. Kikuchi, N. Kanzaki, and K. Futai. 2004. Genet distribution of sporocarps and ectomycorrhizas of *Suillus pictus* in a Japanese white pine plantation. *New Phytol* 164:527–541.

Högberg, M.N., and P. Högberg. 2002. Extramatrical ectomycorrhizal mycelium contributes one-third of microbial biomass and produces, together with associated roots, half the dissolved organic carbon in a forest soil. *New Phytol* 154:791–795.

Intergovernmental Panel on Climate Change (IPCC), Contribution of Working Group 1. 2001. *Climate Change 2001: The Scientific Basis*. (J.T. Houghton, Y. Ding, D.J. Griggs, M. Noguer, P.J. van der Linden, and D. Xiaosu, editors) Cambridge University Press, U.K.

Johnson, N.C., P.J. Copeland, R.K. Crookston, and F.L. Pfleger. 1992. Mycorrhizae: possible explanation for yield decline with continuous corn and soybean. *Agron J* 84:387–390.

Johnson, N.C., D.L. Rowland, L. Corkidi, L. Egerton-Warburton, and E.B. Allen. 2003a. Nitrogen enrichment alters mycorrhizal allocation at five mesic to semiarid grasslands. *Ecology* 84.1895–1908.

Johnson, N.C., J. Wolf, and G.W. Koch. 2003b. Interactions among mycorrhizae, atmospheric CO_2 and soil N impact plant community composition. *Ecol Lett* 6:532–540.

Kapulnik, Y., and D.D. Douds, Jr (eds) 2000. *Arbuscular Mycorrhizas: Physiology and Function*. Kluwer Academic Publishers, London.

Kendrick, B. 2000. *The Fifth Kingdom*, third edition. Focus Publishing, Newburyport, MA.

Kiers, E.T., S.A. West, and R.F. Denison. 2002. Mediating mutualisms: farm management practices and evolutionary changes in symbiont co-operation. *J Appl Ecol* 39:745–754.

Klironomos, J.N. 2002. Feedback with soil biota contributes to plant rarity and invasiveness in communities. *Nature* 417:67–70.

Klironomos, J.N. 2003. Variation in plant response to native and exotic arbuscular mycorrhizal fungi. *Ecology* 84:2292–2301.

Klironomos, J.N., M.F. Allen, M.C. Rillig, J. Piotrowski, S. Makvandi-Nejad, B.E. Wolfe, and J.R. Powell. 2005. Abrupt rise in atmospheric CO_2 overestimates community response in a model plant-soil system. *Nature* 433:621–624.

Klironomos, J.N., and M.M. Hart. 2001. Animal nitrogen swap for plant carbon. *Nature* 410:651–652.

Koide, R.T., and T. Wu. 2003. Ectomycorrhizas and retarded decomposition in a *Pinus resinosa* plantation. *New Phytol* 158:401–409.

Langley, J.A., and B.A. Hungate. 2003. Mycorrhizal controls on belowground litter quality. *Ecology* 84:2302–2312.

Leake, J.R., D.P. Donnelly, and L. Boddy. 2002. Interactions between ecto-mycorrhizal and saprotrophic fungi. In: van der Heijden, M.G.A., and Sanders, I. (eds) *Mycorrhizal Ecology*. Springer, Berlin, Heidelberg, New York. 345–374 pp.

Leake, J.R. 1994. The biology of myco-heterotrophic ("saprophytic") plants. *New Phytol* 127:171–216.

LePage, B.A., R.S. Currah, R.A. Stockey, and G.W. Rothwell. 1997. Fossil ectomycorrhizae from the middle Eocene. *Am J Bot* 84:410–412.

Lilleskov, E.A., T.J. Fahey, and G.M. Lovett. 2001. Ectomycorrhizal fungal aboveground community change over an atmospheric nitrogen deposition gradient. *Ecol Appl* 11:397–410.

Lindahl, B., A.F.S. Taylor, and R.D. Finlay. 2002. Defining nutritional constraints on carbon cycling in boreal forests – towards a less phytocentric perspective. *Plant Soil* 242:123–135.

Lindahl, B., J. Stenlid, and R. Finlay. 2001. Effects of resource availability on the mycelial interactions and P-32 transfer between a saprotrophic and a ectomycorrhizal fungus in soil microcosms. *FEMS Microbiol Ecol* 38:43–52.

Linderman, R.G. 1988. Mycorrhizal interactions with the rhizosphere microflora: the mycorrhizosphere effect. *Phytopathology* 78:366–371.

Linderman, R.G. 2000. Effects of mycorrhizas on plant tolerance to diseases. In: Kapulnick, Y., and Douds, D.D. Jr (eds) *Arbuscular Mycorrhizas: Physiology and Function*, Kluwer Academic Press. 345–366 pp.

Marler, M.J., C.A. Zabinski, and R.M. Callaway. 1999. Mycorrhizae indirectly enhance competitive effects of an invasive forb on a native bunchgrass. *Ecology* 80: 1180–1186.

Miller, R.M., and J.D. Jastrow. 2000. Mycorrhizal fungi influence soil structure. In: Kapulnik, Y., and Douds, D.D. Jr (eds) *Arbuscular Mycorrhizas: Physiology and Function*. Kluwer Academic Publishers, London. 3–18 pp.

Miller, R.M., D.R. Reinhardt, and J.D. Jastrow. 1995. External hyphal production of vesicular-arbuscular mycorrhizal fungi in pasture and tallgrass prairie communities. *Oecologia*, 103:17–23.

Mosse, B. (1973) Advances in the study of vesicular-arbuscular mycorrhizas. *Ann Rev Phytopath* 11:171–196.

Newsham, K.K., A.H. Fitter, and A.R. Watkinson. 1995. Multi-functionality and biodiversity in arbuscular mycorrhizas. *Trends Ecol Evol* 10:407–411.

Nilsson, L.-O., and H. Wallander. 2003. The production of external mycelium by ectomycorrhizal fungi in a Norway spruce forest was reduced in response to nitrogen fertilization. *New Phytol* 158:409–416.

Olsson, P.A., I. Thingstrup, I. Jakobsen, and E. Bååth. 1999. Estimation of the biomass of arbuscular mycorrhizal fungi in a linseed field. *Soil Biol Biochem* 31:1879–1887.

Peng, S., D.M. Eissenstat, J.H. Graham, K. Williams, and N.C. Hodge. 1993. Growth depression in mycorrhizal citrus at high-phosphorus supply. *Plant Physiol* 101:1063–1071.

Pirozynski, K.A., and D.W. Malloch. 1975. The origin of land plants: a matter of mycotrophism. *Biosystems* 6:153–164.

Querejeta, J.I., L.M. Egerton-Warburton, and M.F. Allen. 2003. Direct nocturnal water transfer from oaks to their mycorrhizal symbionts during severe soil drying. *Oecologia* 134:55–64.

Rasanayagam, S., and P. Jeffries. 1992. Production of acid is responsible for antibiosis by some ectomycorrhizal fungi. *Mycol Res* 96:971–976.

Read, D.J. 1991. Mycorrhizas in ecosystems. *Experientia* 47:376–391.

Read, D.J., and R. Boyd. 1986. Water relations of mycorrhizal fungi and their host plants. In: Ayres, P.G., and Boddy, L. (eds) *Water, Fungi and Plants.* Cambridge University Press, Cambridge, 287–303 pp.

Read, D.J., and J. Perez-Moreno. 2003. Mycorrhizas and nutrient cycling in ecosystems – a journey towards relevance? *New Phytol* 157:475–492.

Redecker, D., R. Kodner, and L.E. Graham. 2000. Glomalean fungi from the Ordovician. *Science* 289:1920–1921.

Rillig, M.C., K.K Treseder, and M.F. Allen. 2002. Global change and mycorrhizal fungi. In: van der Heijden, M.G.A., and Sanders, I. (eds) Mycorrhizal Ecology. Springer, New York. 135–160 pp.

Rillig, M.C., S.F. Wright, M.F. Allen, and C.B. Field. 1999. Rise in carbon dioxide changes soil structure. *Nature* 400:628.

Rillig, M.C., S.F. Wright, K.A. Nichols, W.F. Schmidt, and M.S. Torn. 2001. Large contribution of arbuscular mycorrhizal fungi to soil carbon pools in tropical forest soils. *Plant Soil* 233:167–177.

Ryan, M.H., and J.H. Graham. 2002. Is there a role for arbuscular mycorrhizal fungi in production agriculture? *Plant Soil* 244:263–271.

Schultz, P.A., R.M. Miller, J.D. Jastrow, C.V. Rivetta, and J.D. Bever. 2001. Evidence of a mycorrhizal mechanism for the adaptation of *Andropogon gerardii* (Poaceae) to high- and low-nutrient prairies. *Am J Bot* 88:1650–1656.

Setälä, H. 1995. Growth of birch and pine-seedlings in relation to grazing by soil fauna on ectomycorrhizal fungi. *Ecology* 76:1844–1851.

Shi, L., M. Guttenberger, I. Kottke, and R. Hampp. 2002. The effect of drought on the mycorrhizas of beech (*Fagus sylvaticus* L.): changes in community structure, and the content of carbohydrates and nitrogen storage bodies of fungi. *Mycorrhiza* 12:303–311.

Simard, S.W., D. Durall, and M. Jones. 2002. Carbon and nutrient fluxes within and between mycorrhizal plants. In: van der Heijden, M.G.A., and Sanders, I. (eds) *Mycorrhizal Ecology.* Springer, Berlin, Heidelberg, New York. 33–74 pp.

Simard, S.W., D.A. Perry, M.D. Jones, D.D Myrold, D.M. Durall, and R. Molina. 1997. Net transfer of carbon between ectomycorrhizal tree species in the field. *Nature* 388:579–582.

Smith, S.E., and D.J. Read. 1997. *Mycorrhizal Symbiosis.* Academic Press, New York.

Söderberg, K.H., P.A. Olsson, and E. Bååth. 2002. Structure and activity of the bacterial community in the rhizosphere of different plant species and the effect of arbuscular mycorrhizal colonisation. *FEMS Microbiol Ecol* 40:223–231.

Staddon, P.L., A.H. Fitter, and J.D. Graves. 1999. Effect of elevated atmospheric CO_2 on mycorrhizal colonization, external mycorrhizal hyphal production and phosphorus inflow in *Plantago lanceolata* and *Trifolium repens* in association with the arbuscular mycorrhizal fungus *Glomus mosseae. Global Change Biol* 5:347–358.

Stahl, P.D., and W.K. Smith. 1984. Effects of different geographic isolates of *Glomus* on the water relations of *Agropyron smithii. Mycologia* 76:261–267.

Stubblefield, S.P., T.N. Taylor, and J.M. Trappe. 1987. Fossil mycorrhizae: a case for symbiosis. *Science* 237:59–60.

Swaty, R.S., R. Deckert, T.G. Whitham, and C.A. Gehring. 2004. Ectomycorrhizal abundance and community composition shifts with drought: predictions from tree rings. *Ecology,* 85:1072–1084.

Swaty, R.S., C.A. Gehring, M. VanErt, T.C. Theimer, P. Keim, and T.G. Whitham. 1998. Temporal variation in temperature and rainfall differentially affects ectomycorrhizal colonization at two contrasting sites. *New Phytol* 139:733–739.

Treseder, K.K., and M.F. Allen. 2000. Mycorrhizal fungi have a potential role in soil carbon storage under elevated CO_2 and nitrogen deposition. *New Phytol* 147:189–200.

Treseder, K.K., and M.F. Allen. 2002. Direct nitrogen and phosphorus limitation of arbuscular mycorrhizal fungi: a model and field test. *New Phytol* 155:507–515.

van Breemen, N., R. Finlay, U. Lundström, A.G. Jongmans, R. Giesler, and M. Olsson. 2000. Mycorrhizal weathering: a true case of mineral plant nutrition? *Biogeochemistry* 49:53–67.

van der Heijden, M., J.N. Klironomos, M. Ursic, P. Moutoglis, R. Streitwolf-Engle, T. Boller, A. Wiemken, and I.R. Sanders. 1998. Mycorrhizal fungal diversity determines plant biodiversity, ecosystem variability and productivity. *Nature* 396:69–72.

van der Heijden, M.G.A., and I.R. Sanders. (eds) 2002. *Mycorrhizal Ecology*. Springer, New York.

van Tichelen, K.K., and J.V. Colpaert. 2000. Kinetics of phosphate absorption by mycorrhizal and non-mycorrhizal Scots pine seedlings. *Physiol Plant* 110:96–103.

Vitousek, P.M., J.D. Aber, R.W. Howarth, G.E. Likens, P.A. Matson, D.W. Schindler, W. H. Schlesinger, and D.G. Tilman. 1997. Human alteration of the global nitrogen cycle: sources and consequences. *Ecol Appl* 7:737–750.

Walker, R.F., D.R. Geisinger, D.W. Johnson, and J.T. Ball. 1997. Elevated atmospheric CO_2 and soil N fertility effects on growth, mycorrhizal colonization, and xylem potential of juvenile ponderosa pine in a field soil. *Plant Soil* 195:25–36.

Wallander, H. 1995. A new hypothesis to explain allocation of dry matter between mycorrhizal fungi and pine seedlings in relation to nutrient supply. *Plant Soil* 168–169:243–248.

Wallander, H., K. Arnebrant, and A. Dahlberg. 1999. Relationships between fungal uptake of ammonium, fungal growth and nitrogen availability in ectomycorrhizal Pinus sylvestris seedlings. *Mycorrhiza* 8:215–223.

Wallenda, T., and I. Kottke. 1998. Nitrogen deposition and ectomycorrhizas. *New Phytol* 139:169–187.

Wolf, J., N.C. Johnson, D.L. Rowland, and P.B. Reich. 2003. Elevated carbon dioxide and plant species richness impact arbuscular mycorrhizal fungal spore communities. *New Phytol* 157:579–588.

Wright, S.F., M. Franke-Snyder, J.B. Morton, and A. Upadhyaya. 1996. Time-course study and partial characterization of a protein on hyphae of arbuscular mycorrhizal fungi during active colonization of roots. *Plant Soil* 181:193–203.

CHAPTER 5

Soil Rhizosphere Food Webs, Their Stability, and Implications for Soil Processes in Ecosystems

John C. Moore, Kevin McCann, and Peter C. de Ruiter

5.1 INTRODUCTION

The rhizosphere includes plant roots and the surrounding soil that is influenced by plant roots. This definition is more inclusive than traditional definitions that include roots and the soils that adhere to them, by emphasizing that the rhizosphere extends into soils by roots and the actions of root products (Coleman *et al.* 1983, 1996; Van der Putten *et al.* 2001; Moore *et al.* 2003). The traditional definition does more than omit important biological interactions with soil biota, as it perpetuates a framework that views soils as a physical entity, as opposed to a biologically complex and active environment. Each chapter in this book and treatments elsewhere emphasize the diverse and complex nature of these interactions between plants and soil biota within the rhizosphere, and the important roles of soil biota operating within the rhizosphere to plant growth and community dynamics. Our aim is to present a modeling framework that demonstrates that there are generalities in rhizosphere interactions and functions that can be captured and explored mathematically.

The rhizosphere food web we present is compartmentalized into assemblages of organisms in food sub-webs supported by bacteria and their consumers, fungi and their consumers, and the plant roots and their consumers. Three key features that distinguish the organisms within one assemblage from another are that (1) they process different types of energy inputs at different rates; (2) they possess different life-history characteristics; and (3) they occupy different microhabitats. We argue that trophic interactions within the rhizosphere are best studied as a grouping of these sub-webs operating in concert, and also possessing quasi-independent tendencies. There is a solid rationale for this approach (Bender *et al.* 1984; O'Neill *et al.* 1986; Yodzis 1996), particularly when it comes to subsets of species sharing similar and intertwined dynamics, per capita effects, feeding rates, death rates, and growth rates.

We will explore how these three major assemblages are structured and interact in ways that are important to the stability of the food web and thus the persistence of crucial ecosystem functions. We present empirical evidence and theoretical exercises that precipitate changes in the relative sizes and rates of nutrient transfer within these assemblages to demonstrate these points. We also consider how the rhizosphere food web can affect the structure and dynamics of plants, herbivores, and predators living aboveground.

5.2   THE STRATEGY UNDERLYING MATHEMATICALLY CAPTURING THE ESSENCE OF RHIZOSPHERE FUNCTION

As students of biology, we design and are exposed to a variety of representations to convey complex interactions and relationships. We are all acquainted with the first representation, what for a better term can be referred to as the fundamental equation of life:

$$6CO_2 + 6H_2O \leftrightarrow C_6H_{12}O_6 + 6O_2 \tag{5.1}$$

Equation 5.1 encompasses several concepts. It illustrates the conservation of matter, as each side of the equation possesses different molecules but equal numbers of atoms and mass. This conservation is essential as we develop mathematical models depicting rhizosphere interactions. Apart from representing photosynthesis and respiration, Equation 5.1 depicts the interdependence between life and death processes operating at different scales, the autotroph and heterotroph, the interaction between a plant and a herbivore, and the immobilization of inorganic matter into organic matter and the mineralization of organic matter to inorganic matter. Of course, life is not only composed of carbon, water, and hydrogen, and rhizosphere function is particularly important in recycling of nutrients such as N and P. If we add nitrogen to the equation, taking into account that different kinds of organisms have different ratios of N, P, C, and other elements in their biomass (different stoichiometries), a similar set of equations and associated processes emerges, and the interdependence of elements in shaping rates and life processes is evident (Reiners 1986; Sterner and Elser 2001). As with carbon, nitrogen is immobilized into organic matter and mineralized into inorganic matter, but we see an added dimension of a tight coupling of the compartmentalized above- and belowground processes as organisms from within each realm have perfected the biogeochemical pathways to immobilize the inorganic metabolic wastes and excesses of the other.

Students are also familiar with food-web caricatures used to depict trophic interactions (Moore *et al.* 2005). The figures often include a plant, and multiple aboveground herbivores and predators, and if the vignette is of a terrestrial systems, often an arrow links detritus to soils down below the soil surface, to "nutrients" and/or "microbes" (indicating nutrient cycling), followed by an arrow pointing to plant roots (indicating nutrient uptake). The clear emphasis of these depictions is on the aboveground realm, even though the interactions occurring belowground within the rhizosphere may be as or more significant in scope, complexity, and overall importance to the system. In these depictions the aboveground system receives greater attention, if not purely for heuristic reasons, while soils and soil processes are given short shrift stems because of the obscure nature of soil biota and processes.

Finally, the mathematical models we will develop to capture ideas of stoichiometry and trophic interactions (energy transfers) in the rhizosphere represent a third type of caricature. On the one hand, effective models are internally consistent, simple in design and assumption, and thought provoking. On the other hand, they can be devoid of the details that make them biologically interesting and thus lead to biologically counterintuitive results. A good example of the latter is the unstable mathematical representations of mutualisms emerging from theory (Pimm 1982) in contrast to the ubiquitous nature of what appear to be stable symbiotic mutualisms that occur within the rhizosphere that have evolved over time.

To meet the aim of this chapter we will present a mathematical approach that incorporates the reciprocal transfer of nutrients that are essential for plant growth and heterotrophic life depicted in Equation 5.1, and the trophic interactions among organisms above- and belowground, as depicted in the caricatures described above. We will then use this approach to demonstrate that the rhizosphere possesses a distinct trophic structure that is important to mathematical stability, and that human activities can alter the structure that are mathematically unstable and in ways that alter key ecological process.

5.3   RESOURCE FLOW IN THE RHIZOSPHERE

To begin to capture functions within the rhizosphere mathematically, it is important first to note the resource interactions in the rhizosphere that we will include in our models and the currency that we will base them on. More extensive treatments of these interactions can be found in Chapters 2–4. First, significant quantities of photosynthetic products produced by plants are diverted to roots for root growth, which provides a carbon base for the soil species. The rhizosphere is characterized by rapid and prolific root growth, the sloughing of root cells, root death, and the exudation of simple carbon compounds. The size and dynamic of the rhizosphere relative to the aboveground component of plants differ by plant species and ecosystem type. For example, in grasslands, the ratio of shoot to root (S:R) production is roughly 1:1, contrasting sharply with forests, where far more photosynthate is allocated aboveground (Jackson *et al.* 1996), while Arctic tundra is characterized by a rhizosphere that turns over slowly resulting in an accumulation of root materials (Shaver *et al.* 1992). Interestingly, the range in S:R is narrowly conserved between 0.1 and 5 (Farrar *et al.* 2003), significant when contrasted with the range in plant sizes. The reasons offered for the constancy in S:R are centered on the constraints on plant imposed by limitations and invariance in C:N and C:P ratios and the selective pressure to acquire just enough of the soil-based resources to balance aboveground carbon fixation. The constancy in the S:R and the dependence on elemental ratios greatly simplifies and strengthens our ability to generalize any models that we may develop.

The carbon flux into the rhizosphere helps support the rhizosphere food web function that we aim to describe in mathematical form. Detailed studies of the rhizosphere reveal that a growing root can be subdivided into a continuum of zones of activity from the root tip to the crown where different microbial populations have access to a continuous flow of organic substrates derived from the root (Trofymow and Coleman 1982). The root tip represents the first and lowest root zone. It is the site of root growth and is characterized by rapidly dividing cells and secretions or exudates that lubricate the tip as it passes through the soil. The exudates and sloughed root cells provide carbon for bacteria and fungi which in turn immobilize nitrogen and phosphorous. Farther up the root is the region of nutrient exchange, characterized by root hairs and lower rates of exudation. The birth and death of root hairs stimulates additional microbial growth. The upper zones have been characterized as the region of remineralization of nutrients by predators, the region of symbiotic mutualistic relations, and the structural region (Coleman *et al.* 1983). Within each of the zones there is an infusion of carbon into the rhizosphere by plants which stimulate the growth and activity of microbes (Foster 1988; Grayston *et al.* 1996; Bardgett *et al.* 1998; Bringhurst *et al.* 2001) and the Protozoa and invertebrates that feed on them (Lussenhop and Fogel 1991; Parmelee *et al.* 1993).

IDENTIFICATION OF INTERACTIVE FOOD WEBS WITHIN THE RHIZOSPHERE

The mathematical depictions of the food webs within the rhizosphere we present are based on the three types of food web descriptions presented by

Paine (1980) – *connectedness, energy flow,* and *interaction.* Each description builds on the other in terms of the information required to construct them.

Connectedness descriptions are based on observations of the species that are present and their feeding behaviors. Given the minute sizes of soil microbes and invertebrates, most descriptions of the trophic interactions within soils are based on the observations presented in the literature, the gut contents of collected specimens, arena studies in small microcosms, and more recently the concentrations of stable isotopes. *Connectedness* descriptions, including those presented below, are notoriously incomplete, and grossly over-simplify the diversity and complexity of the system being studied.

The *energy flow web* quantifies the amount of energy within organisms and transfer of energy among organisms, though usually represented in terms of C, N, or P. If nutrients or elements are used to represent energy flow webs, as we will do below, a few words of caution are in order. Implicit in this decision is the assumption that material flows (nutrients or elements) and energy are interchangeable for the purpose of describing food web structure and function. Although the energy is indeed contained in C–C bonds, transferred around to various forms, for example ATP and NADPH to do work, the transfer of nutrients is not necessarily tied in a uniform way to transfer of energy. However, variation in the energy and material content of organisms is far less than the variation in flows among organisms, making either useful indices of patterns of flow (energy or matter) through food webs.

The *interaction web* depicts the influences of the dynamics of one group on another. These descriptions have been adopted by several research groups that have attempted to link the structure of soil food webs in relation to the decomposition of organic matter and the mineralization of nutrients (Hendrix *et al.* 1986; Hunt *et al.* 1987; Brussaard *et al.* 1988, 1997; Moore *et al.* 1988; Andrén *et al.* 1990; de Ruiter *et al.* 1993a, b).

DEVELOPMENT OF THE MATHEMATICAL REPRESENTATION OF A RHIZOSPHERE FOOD WEB

We present a description of the rhizosphere food web of the North American shortgrass steppe to introduce our approach. The *connectedness web* defines the model's basic structure, indicating what organisms consume what other organisms or substrates (Figure 5.1). The diagram simplifies the high complexity and diversity of the rhizosphere community by defining the web in terms of functional groups of organisms that share similar prey and predators, feeding modes, life history attributes, and habitat preferences (Moore *et al.* 1988). At the base of the web are plant roots, labile (C:N ratio <30:1) and resistant (C:N

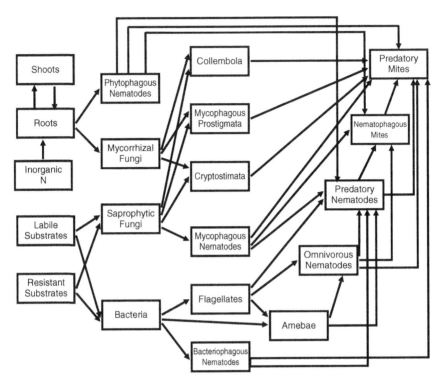

FIGURE 5.1   A combined depiction of the connectedness, energy flux, and interaction food webs of the belowground food web of the North American shortgrass steppe, Nunn, CO, USA. The boxes represent functional groups of species that share food types, feeding modes, habitats, and life history traits. The average annual biomass $(kg\,C\,ha^{-1})$ estimates and physiological attributes of each group are presented in Table 5.1.

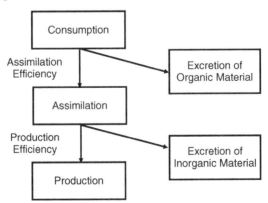

FIGURE 5.2   Scheme for estimating feeding rates of individual predator–prey interactions in terms of consumption, biomass, and the excretion of organic and inorganic matter (after Hunt *et al.* 1987).

TABLE 5.1 Physiological parameter values and average annual population densities ($kg\,C\,ha^{-1}$) for the functional groups of the belowground food web of the shortgrass steppe, Nunn, CO, USA depicted in Figure 5.1 (after Hunt et al. 1987). The functional groups are loosely arranged by trophic level with lower levels at the bottom and upper levels at the top

Functional group	C:N	Turnover rate (yr^{-1})	Assimilation efficiency (%)	Production efficiency (%)	Biomass ($kg\,C\,ha^{-1}$)
Predatory Mites	8	1.84	60	35	0.160
Nematophaous Mites	8	1.84	90	35	0.160
Predatory Nematodes	10	1.60	50	37	1.080
Omnivorous Nematodes	10	4.36	60	37	0.650
Fungivorous Nematodes	10	1.92	38	37	0.410
Bacteriophagous Nematodes	10	2.68	60	37	5.800
Collembola	8	1.84	50	35	0.464
Mycophagous Prostigmata	8	1.84	50	35	1.360
Cryptostigmata	8	1.20	50	35	1.680
Amebae	7	6.00	95	40	3.780
Flagellates	7	6.00	95	40	0.160
Phytophagous Nematodes	10	1.08	25	37	2.900
AM-Mycorrhizal Fungi	10	1.20	100	30	7.000
Saprobic Fungi	10	2.00	100	30	63.000
Bacteria	4	1.20	100	30	304.000
Detritus	10	0.00	100	100	3000.000
Roots	10	1.00	100	100	300.000

ratio > 30:1) forms of detritus, and an inorganic nitrogen source. The plant roots and their products and detritus are utilized by a host of microbes and invertebrates, terminating with predatory mites.

The *energy flow web* expresses food web structure in quantitative measures, that is population sizes (biomass) and feeding rates. The estimates of flow can be derived indirectly using estimates of population sizes, turnover rates, consumption rates, prey preferences, and energy conversion parameters (Table 5.1, see de O'Neill 1969; Hunt et al. 1987; de Ruiter et al. 1993b). In our example food web, feeding rates were estimated using the procedures (Figure 5.2) presented by Hunt et al. (1987). Consumed matter is divided into a fraction that is immobilized into consumer biomass (assimilation) and a fraction that is returned to the environment as feces, orts, and unconsumed prey, and of the assimilated fraction, material that is incorporated into new biomass (production) and material that is mineralized as inorganic material. The estimates begin with top predators with the assumptions that the amount of material required to maintain the predator's steady state biomass must equal the sum of its steady state biomass and loss due to death divided by its ecological efficiency:

$$F = (D_{nat}B + P)/e_{ass}e_{prod} \qquad (5.2)$$

where F is the feeding rate (biomass time^{-1}), D_{nat} is the specific death rate, or turnover rate (time^{-1}) of the consumer, B (biomass) is the population size of the consumer, P is the death rate to predators (biomass time^{-1}), and e_{ass} and e_{prod} are the assimilation (%) and production (%) efficiencies, respectively.

Hence, to estimate the flux F for a top predator, assume that the death due to predator is zero, and the inverse of its life span represents a first approximation of its specific death rate, D_{nat}. The biomass, B, of the predator can be approximated from field collections, while the energetic efficiencies, e_{ass} and e_{prod}, are obtained from laboratory studies of field taxa or published values for similar taxa. For predators that consume multiple prey types the fluxes are weighted by the feeding preferences of the predator for the respective prey. The estimation procedure moves downward through the prey to the basal resources with fluxes to each prey taking into account the biomass lost to predation. A dynamic version can be constructed by taking into account changes in the biomasses over an interval of time t, that is adding $\Delta B/t$ to the numerator of Equation 5.2.

The *interaction web* emphasizes the strengths of the interactions among the functional groups. The idea of interaction strengths can be simply described as the effect of a change in biomass (population) of one organism (e.g., prey or predator) on the change in biomass (population) of another organism of interest. Interaction strengths can be examined for all pairs of interacting organisms in a food web, in order to discern, for example, whether prey availability or predator abundance is the dominant control over a particular organism's population size (or biomass). The interaction strengths have important implications for whether there is "top-down" or "bottom-up" control over particular trophic groups at various locations within the food web. Moore et al. (1993) and de Ruiter et al. (1995) developed a means to estimate interaction strengths from the energy flow web (Hunt et al. 1987) and the differential equations used to describe the dynamics of each functional group.

The interaction strengths are defined by the partial derivatives of the equations describing the growth and dynamics of the functional groups at or near equilibrium (May 1973):

$$\alpha_{ij} = [\delta(dX_i/dt)/\delta X_j]^* \qquad (5.3)$$

where α_{ij} refers to the interaction strength (per unit biomass effect) of functional group i on functional group j, and the asterisk indicated that the partial derivatives are evaluated near equilibrium. Interaction strengths can be derived directly from equations used to model the population dynamics of the functional groups if the population densities of the functional groups and the feeding rates (Equation 5.2) are known. The key assumptions behind the estimation procedure are as follows: (1) the equilibria of the functional groups represented in the differential equations can be approximated with long-term seasonal field averages of the functional groups (e.g., $X_i^* = B_i$); and (2) the consumption terms in the differential equations for prey i to predator j can be approximated from the flux rates described in Equation 5.2, i.e., $F_{ij} = c_{ij}X_i^*X_j^*$, where c_{ij} is the consumption coefficient of functional group j on functional group i. Hence, if derived from rate equations based on Lokta–Volterra rate equations, the interaction strengths are $\alpha_{ij} = -F_{ij}/B_j$ for the per capita effect of predator j on prey i, and $\alpha_{ji} = a_j p_j F_{ij}/B_i$ for prey i on predator j (Moore et al. 1993; de Ruiter et al. 1994, 1995). The diagonal elements of the matrix cannot be derived from field data or estimates of energy fluxes, but can be scaled to the specific death rates (de Ruiter et al. 1995) or can be set at levels that ensure stability (Neutel et al. 2002) depending on your aims.

PATTERNS IN STRUCTURE

The connectedness and energy flux descriptions reveal two patterns in the distribution of energy, nutrients, and biomass within the system that are important to its stability. The first pattern deals with the flow of energy from roots and detritus to top predators. The food web that develops within the rhizosphere is complex (Figure 5.1), consisting of multiple assemblages of species that originate directly from roots and root by-products (Hunt et al. 1987). The system possesses three distinct pathways or energy channels (Table 5.2) originating from living plant roots, resistant detritus through fungi, and labile detritus through bacteria (Coleman 1976; Coleman et al. 1983; Hunt et al. 1987; Moore and Hunt 1988). Moore et al. (1988) described these assemblages as the root, bacterial, and fungal energy channels, the organisms within which share distinct physiological and behavioral attributes (Table 5.3) in terms of their resource utilization that lend themselves to this type of compartmentalization (Schoener 1974). Root-feeding insects and nematodes, root

TABLE 5.2   Estimates of the percentage of energy that each functional group derives from bacteria, fungi, and roots from within the belowground food web of the shortgrass steppe, Nunn, CO, USA (Moore and Hunt 1988). The estimates were derived from the feeding rates as calculated by Equation 5.2

| | | Energy channel | |
Functional group	Bacteria	fungi	Root
Protozoa			
Flagellates	100	0	0
Amebae	100	0	0
Ciliates	100	0	0
Nematodes			
Phytophagous Nematodes	0	0	100
Mycophaogous Nematodes	0	90	10
Omnivores Nematodes	100	0	0
Bacteriophagous Nematodes	100	0	0
Predatory Nematodes	68.67	3.50	27.83
Microarthropods			
Collembola	0	90	10
Cryptostigmata	0	90	10
Mycophagous Prostigmata	0	90	10
Nematophagous Mites	66.70	3.78	29.52
Predatory Mites	39.54	38.56	21.91

TABLE 5.3   Habitat use, life history characteristics, and energetic efficiencies of broad classifications of organisms encountered in belowground food webs (Hunt *et al.* 1987, Coleman 1996, and Moore and de Ruiter 1997). These attributes along with the food preferences serve as the basis of the compartmentalization of trophic interactions along the principal niche axes of food, habitat, and time (Schoener 1974)

Taxon habitat	Bacteria water/ surfaces	Fungi free/ surfaces	Protozoa water/ surfaces	Microbivorous nematodes water films/ surfaces	Collembola free	Mites free
Minimum Generation Time (h)	0.5	4–8	2–4	120	720	720
Turnover Time $(season^{-1})$	2–3	0.75	10	2–4	2–3	2–3
Assimilation Efficiency (%)	100*	100*	95	38–60	50	30–90
Production Efficiency (%)	40–50	40–50	40	37	35	35–40

* The Assimilation Efficiencies of bacteria and fungi are 100% given that microbes absorb materials across their membranes as opposed to ingesting or engulfing prey or materials.

pathogens, and microbes that engage in symbiotic relationships with plant roots (e.g., mycorrhizal fungi, *Rhizobium*, *Frankia*) form the base of the root energy channel. The bacterial energy channel consists of saprophytic bacteria, protozoa, nematodes, and a few arthropods. The fungal energy channel largely consists of saprophytic fungi, nematodes, and arthropods. Soil bacteria compose most of the microbial biomass in the rhizosphere, are aquatic organisms, and are more efficient in using the more labile root exudates than saprophytic fungi (Curl and Truelove 1986). In contrast, fungi are more adapted to utilize more resistant root cell components and substrates than are bacteria. Moreover, fungi and their consumers occupy air filled pore spaces and water films, and possess longer generation times. Nutrients within each channel are processed at different rates given the differences in the recalcitrance of the materials that bacteria and fungi utilize and the physiologies of the fauna within each channel. Coleman *et al.* (1983) recognized these differences and referred to what became known as the bacterial energy channel as a "fast cycle," while what came to be known as the fungal energy channel represented a "slow cycle." Importantly, mathematical representations of the type of compartmentalized architecture whose subsystems differ in dynamic properties as described above have been shown to be more dynamically stable than random constructs possessing the same diversity (number of groups) and complexity (number of linkages among groups) (May 1972, 1973; Moore and Hunt 1988; Yodzis 1988).

The second pattern deals with the distribution of biomass with increased trophic level. The systems we have described to date possess steep trophic pyramids of biomass and feeding rates (Table 5.1). For example, bacteria and fungi account for upward of 95 percent of consumer biomass, leaving less than 5 percent for the protozoa and invertebrates occupying the upper trophic positions (Hunt *et al.* 1987; de Ruiter *et al.* 1993b). Recent work has demonstrated that the pyramidal structure is more stable than alternative structures, for example inverted pyramids, with higher biomass maintained at upper trophic levels (Moore and de Ruiter 2000; Neutel *et al.* 2002; Moore *et al.* 2003).

Beyond the distribution of biomass and compartmentalized energy flow patterns noted above, the interaction web possesses an asymmetric pattern in the pairwise interaction strengths (i.e., the positive effects of prey on their predators and the negative effects of predators on their prey) that is dependent on trophic level. At the lower trophic levels, predators have relatively strong negative effects on prey, and at higher trophic levels, prey have relatively strong positive effects on predators. This patterning in the interaction strengths is linked with food web stability, as soil food webs that possess this pattern are more stable than those without the pattern (de Ruiter *et al.* 1995; Moore *et al.* 1996; de Ruiter *et al.* 1998). When the ratios of the absolute values of the paired interaction strengths of predator on prey and prey on predator are plotted by trophic position, the slopes of the lines for the whole web and those for the different energy channels are similar. This repeating of pattern in a fractal-like invariant manner within each of the energy channels (Figure 5.3) reinforces the evidence that the system is compartmentalized along energy utilization (Moore and de Ruiter 1997). In retrospect, these results are not surprising given that the patterning of interaction strength is also intimately linked with the distribution of biomass and the feeding rates discussed above, given that both are components of the estimates of interaction strength. Redistributing interaction strengths in the analysis concomitantly redistributes biomass and feeding rates.

SHIFTS AMONG ENERGY CHANNELS WITHIN THE RHIZOSPHERE

The relative dominance of the root, bacterial, and fungal energy channels is important to key processes and stability (Moore *et al.* 2005). Studies of

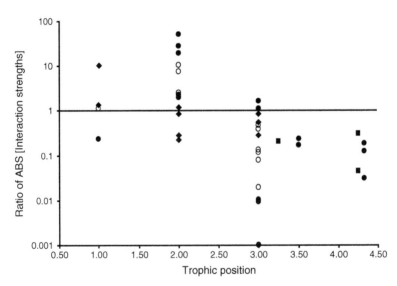

FIGURE 5.3 The asymmetry in the interaction strengths with increased trophic position for the belowground food web of the shortgrass steppe that was observed in Figure 5.3 can also be found within each energy channel (bacterial channel ●, fungal channel O, and root channel ◆). Each symbol represents the ratio of the pairwise interaction strengths between a consumer and a resource. Prior to taking the ratio, the interaction strengths were standardized by dividing each α_{ij} and α_{ji} by the average interaction strengths for all α_{ij} and α_{ji}, respectively. Predators (■) were considered separately as significant proportions of their energy are obtained from more than one energy channel (Table 5.2). Figure adapted from Moore and de Ruiter (1997).

grasslands, forests, arctic tundra, and agricultural systems indicate that the linkages among the energy channels tend to be weak at the trophic levels occupied by roots, bacteria, and fungi, and strongest at the trophic levels occupied by predatory mites. The strength of the linkages among energy channels and the dominance of a given energy channel varies by the type of ecosystem, can change with disturbance, and affects nutrient turnover rates (Figure 5.4). The fungal energy channel tends to be more dominant in systems where the ratio of carbon to nitrogen of detritus is high (e.g., forests, no-till agriculture) while the bacterial channel is more dominant in systems with narrow ratios (e.g., grasslands, conventional tillage agriculture). Regardless of the relative dominance of either channel, disturbances that either add labile nitrogen through fertilization, disrupt soil aggregates, or remove vegetation have been shown to induce shifts in the linkage between the fungal and bacterial energy channels that favor the bacterial channel (Figure 5.4). Accelerated rates of decomposition, increases in the rates of nitrogen mineralization, and increases in nitrogen loss have been associated with an increase in the dominance of the bacteria energy channel (Hendrix

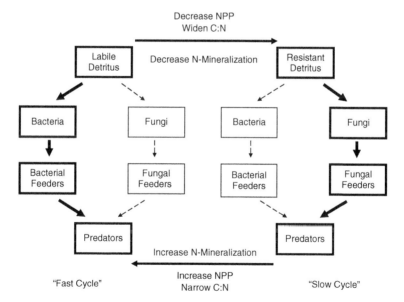

FIGURE 5.4 Simplified bacterial and fungal energy channels of a rhizosphere food web. The bacterial energy channel represents the "fast cycle" due to the higher turnover rates of bacteria and their consumers relative to the fungi and their consumers. The fungal energy channel represents the "slow cycle." Changes in the C:N ratio of the detritus, NPP, or rates of N mineralization have been associated with shifts in the relative dominance of one channel to the other (adapted from Moore *et al.* 2003).

et al. 1986; Andrén *et al.* 1990; Moore and de Ruiter 1991; Doles 2000; Moore *et al.* 2005).

Mathematical representations of these shifts indicate that stability of the rhizosphere food web may ultimately be affected. Working with a simplified representation of the bacterial and fungal energy channels, we find that the coupled pathways are more stable than either pathway operating alone under the same level of energy input, and that shifts in the relative activity of either pathway affects stability (Figure 5.5). We present a model that was originally designed to study trophic cascades, but the structure and outcomes are germane to this discussion (Moore *et al.* 2004, 2005). The model assumes

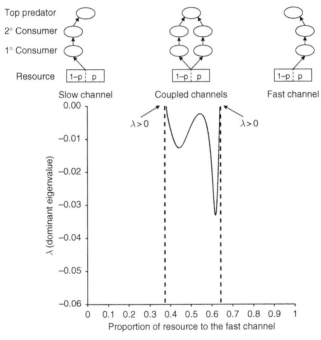

FIGURE 5.5 The relationship of the dominant eigenvalues of a series of models possessing two parallel food chains drawing energy from a heterogeneous source and linked by a common predator (modified from Moore *et al.* 2004). The food chains differ in the rates in which they process and turnover energy, i.e. fast (right) and slow (left) pathways, as encountered with the bacterial and fungal pathways depicted in Figure 5.4. The x-axis represents the proportion of resource passing through the fast channel (p), while the y-axis represents the dominant eigenvalue (λ) for the system for each partitioning of resource. The total amount of energy passing through each system is the same for each partitioning of resource. The dashed vertical lines represent unstable transitions. Note that the most stable configuration occurs when the two pathways are coupled, and the system is unstable if most of the energy passes through the fast pathway ($\lambda > 0$).

that the first level includes a pool of resources (either detritus or roots or both) that supports two parallel food chains linked together by a top predator in the manner presented by Post *et al.* (2000). For our purposes assume that the pathways represent the fast-cycling bacterial energy channel and the slow-cycling fungal energy channel, achieved by varying the physiologies of their respective organisms and the rates that they turnover (see Table 5.1). With the differential equations that describe the interactions structured and parameterized along the lines of McCann *et al.* (1998), we find that the linked pathways were stable and less prone to oscillation when the top predators received from ∼35 to 65 percent of their energy from either pathway (the range where λ is negative), rather than 100 percent from a single pathway. In other words, shifts in the relative activity of one pathway to another may be inherent to the system (de Ruiter *et al.* 2005), but extreme shifts, particularly those that favor the fast-cycling bacterial pathway, may lead to instabilities.

MUTUALISMS WITHIN THE RHIZOSPHERE, AND LINKS TO ABOVEGROUND FOOD WEBS VIA PLANTS

The trophic interactions within the rhizosphere, particularly symbiotic mutualisms, affect plant growth and community structure aboveground. Direct symbiotic mutualisms (*sensu* Boucher *et al.* 1982) involving reciprocal transfers of limiting nutrients between plants and microbes are a prominent feature of rhizospheres of plant communities in terrestrial ecosystems. Studies of primary and secondary succession reveal strong correlations between symbiotic mutualisms, nutrient dynamics, plant growth, and community structure (Reeves *et al.* 1979; Vitousek *et al.* 1987; Vitousek and Walker 1989; Wall and Moore 1999; Moore *et al.* 2003).

Several approaches to modeling symbiotic interactions have been undertaken to explain the aforementioned linkages between the growth responses of the host and the symbiont (Johnson *et al.* 2006). Swartz and Hoeksema (1998) presented an adaptation of market models used in economics and trade that builds on the notion of reciprocal transfers of limiting resources. Under this supposition, the excess resource for one of the partners is the limiting resource for the other, and vice versa. The trading of the excess resources allows for potential increases in the carry capacities of both partners. The parallels in the economic models to symbiotic relationships within the rhizosphere are clear as plants often serve as the host and "trade" carbon or refugia with the symbiont (e.g., mycorrhizal fungi, *Rhizobium*, *Frankia*), in exchange for a plant-limiting nutrient, usually nitrogen or phosphorous. Missing from the conceptual market modeling approach is the explicit treatment of the mechanisms that precipitate the actual trade. Hunt *et al.* (1987) and Moore and

Hunt (1988) treated mycorrhizal interactions as simple trophic interactions, in as much as mycorrhizal fungi served as predators and plants served as prey when modeling carbon, and with the roles reversed when modeling nitrogen. A simple Holling Type I (constant) functional response (sensu Holling 1959) was used to model the interactions under the assumption of mass balance in the early formulations, with later treatments including a more realistic Holling Type II (saturation) functional response to describe the transfers (Moore *et al.* 2003).

While modeling the symbioses as special forms of trophic interactions makes sense, more sophisticated applications are needed if we are to apply the economic models to nutrient exchanges or were to account for the recent revelations of molecular and chemical signaling between plants and microbes that may be involved in nutrient exchanges (Philips *et al.* 2003). Signaling may produce market model outcomes; however, if chemical signals from either the host or symbiont served to up-regulate or down-regulate gene expression in a way that hastened or retarded nutrient release or exchange, it is unlikely that simple Holling Type I, II, or even III functional responses would capture these processes given the complexities of the feedbacks involved.

IMPACTS OF THE RHIZOSPHERE ON PLANT GROWTH AND COMMUNITY DEVELOPMENT

The empirical evidence on mutualism, coupled with the earlier discussion of nutrient enrichment afforded by the trophic interactions within the energy channels, illustrate clear connections between soil biota and plant growth and community structure (Figure 5.6a). Our recommendation for greater sophistication notwithstanding, our models at this point were formulated using differential equations with Holling Type II functional responses describing attack rates and rates of nutrient uptake (Holling 1959) and parameterized after Hastings and Powell (1991). Nevertheless, they suggest that these interactions may influence the overall system's dynamic stability as well (Moore *et al.* 2003). The below- and aboveground components of the system are illustrated in Figure 5.6a and b, and the modeled biomasses of plants, herbivores, and predators resulting from shifting nutrient availability belowground are illustrated in Figure 5.6c. The aboveground component is represented as a simple food chain which consists of a plant that is consumed by a herbivore which in turn is consumed by a predator. The belowground component includes a pool of nitrogen, microbes, and their predators. The growth rate of the plant is dependent on its uptake of nitrogen by plant roots, which is governed by a Type II saturating functional response (Figure 5.6b).

Using these models, we explored how dynamics of the aboveground system are influenced by supply and uptake of nitrogen from belowground. We

(a)

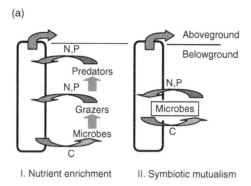

I. Nutrient enrichment   II. Symbiotic mutualism

(b)

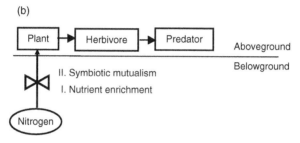

FIGURE 5.6 (a) Conceptual diagram of how soil fauna might influence aboveground dynamics through nutrient enrichment (I) (Clarholm 1985; Ingham *et al.* 1985) and by altering symbiotic interactions (II). The single pathway depicted in the nutrient enrichment model can be split into the bacterial and fungal energy channels presented in Figure 5.4. Rhizodeposition includes both labile and resistant forms of carbon (detritus). Symbiotic interactions include mycorrhizal fungus, *Frankia*, and Rhizobiaceae bacteria. (b) Simple food web model that links an aboveground food chain that includes a plant, a herbivore, and a predator with a belowground food web through a nitrogen source (N). The aboveground food web used the same structure and parameterization as Hastings and Powell (1991). The uptake of nitrogen and all other trophic interactions were modelled using a Type II functional response. The nutrient enrichment model (I) was simulated by altering the rate of input of nitrogen into the nitrogen pool. We simulated the altering of symbiotic relations (II) by changing the half saturation coefficient associated with the uptake of nitrogen by the plant. (c) Phase-space of the biomass for plants vs. herbivores (lower) and predators vs. herbivores (upper) for a plant → herbivore → predator food chain coupled to a decomposer subsystem linked through the rhizosphere as depicted in panels A and B (after Moore *et al.* 2003). The models included Type II functional responses for all the trophic interactions, including the uptake of nitrogen from the decomposer subsystem through the rhizosphere to the plant. Plant productivity was regulated by nitrogen availability (a–c) through rhizosphere activity by altering the saturation constant, within the Type II functional response or the size of the nitrogen pool. The plant → herbivore → predator portion was initialized using the parameter selection from Hastings and Powell (1991), to produce the "teacup" phase-space (c).

(c)

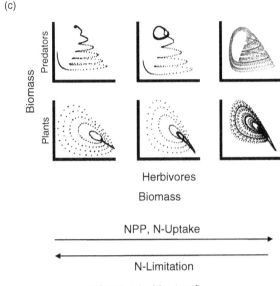

FIGURE 5.6 (*Continued*)

simulated shifts that occur between the bacterial and fungal energy channels by increasing the input of nitrogen into the N-pool, under the assumption that the trophic interactions within each energy channel within the rhizosphere affects the pool size alone and not basic plant physiology related to the rate of uptake of nitrogen. We simulated altering a symbiotic interaction by altering the half saturation parameter in the functional response describing the uptake of nitrogen by the plant, under the assumption that the symbiosis is a physiological link that if altered would directly affect the uptake rate of nitrogen. We recognize that increases in the N-pool and availability of N can negatively influence the infection rates and effectiveness of mycorrhizae, but made no attempt to incorporate this; hence when we increase N-uptake through either mechanism in the model we are representing the net difference in N-availability in uptake by mycorrhizae and the N-pool directly through roots. Nonetheless, both mechanisms yielded similar results (Figure 5.6c).

We present the results of the model runs (1000 time steps) in Figure 5.6c in the form of phase spaces of the biomasses of plants and herbivores (lower tier of graphs) and herbivores and predators (upper tier of graphs) along a gradient of productivity (NPP) induced by increasing the rate of nitrogen uptake. High NPP corresponds to a high uptake rate of N and a low level of N-limitation for the plant. Each of the points within the graphs represent the biomasses of prey and predator at a given time step. When an equilibrium or limit cycle is reached for both prey and predator, the points plot on top of one another with each successive time step giving the appearance of a single point or a loop-like pattern, respectively. Chaotic dynamics is typified by a series of unique trajectories operating within clearly defined bounds.

At the high levels of nitrogen input, the dynamics of the aboveground compartment resembled the results for the three species food chain those presented by Hastings and Powell (1991), as the phase space of the predator and herbivore reveals the familiar chaotic dynamics with a high denisity of unique points within an attractor with boundaries that resembles a upside-down "teacup." We could attenuate the dynamics from chaos to a stable equilibrium by lowering the rate of input of nitrogen into the N-pool, by lowering the uptake of nitrogen from the N-pool, or by lowering the half-saturation parameter in the functional response. At lower levels of nitrogen uptake, the model quickly settles into a stable limit cycle and then to an equilibrium. At these lower levels of nitrogen uptake, the biomasses of prey and predators repeat themselves at successive time steps, as seen by the low density of unique points in the graphs of the phase space.

This modeling exercise indicates that the dynamics of the aboveground system is sensitive to the supply rate and uptake rate of nitrogen by the plant. An increase in the uptake of limiting nutrients by the plant led to greater productivity, and in line with predictions of Rosenzweig (1971), larger oscillations, and/or chaos. A limitation in nutrients yielded limit cycles and convergence to single equilibrium states. While the results are consistent with the conventional wisdom that dynamics is a function of productivity (Rosenzweig 1971; May 1976; Moore *et al.* 1993) and more specifically nutrient supply rates (DeAngelis 1992) they place the more general outcomes in the context of the rhizosphere by providing two specific mechanisms, nutrient enrichment, and altered symbiosis.

5.4 DISCUSSION AND CONCLUSIONS

Each chapter in this book and treatments elsewhere emphasize the important roles of soil biota operating within the rhizosphere to plant growth and community dynamics. Yet, the dated notions of soils being reservoirs of plant-limiting nutrients are still pervasive in introductory texts, and to some extent in our treatment of food webs, trophic dynamics and plant community development.

We have argued that the rhizosphere is best studied as a grouping of sub-webs not only operating in concert, but also possessing quasi-independent tendencies. There is a solid rationale for this approach (Bender *et al.* 1984; O'Neill *et al.* 1986; Yodzis 1996), particularly when it comes to subsets of species sharing similar and intertwined dynamics, per capita effects, and rates. The rhizosphere food web we present is compartmentalized into assemblages of organisms based on bacteria and their consumers, fungi and their consumers, and the plant roots and their consumers. The key feature that distinguishes the assemblages is that they process different types of energy inputs at different rates, that they possess different life histories, and that they occupy different microhabitats. Stepping back and viewed from a more abstract perspective, these attributes are aligned with three principal niche axes of time and habitat, respectively (Schoener 1974), leading to a niche-compartmentalized rhizosphere. We see evidence of the compartmentalization in our descriptions and in our tracking of the food webs following disturbances (Coleman *et al.* 1983; Hendrix *et al.* 1986; Hunt *et al.* 1987; Moore *et al.* 1988; Moore and Hunt 1988; Andrén *et al.* 1990; Moore and de Ruiter 1991; Beare *et al.* 1995). The organisms within the bacterial energy channel and those within the fungal energy channel respond to disturbances to differing degrees as units rather than in a haphazard manner, affecting not only structure but function as well (see Figure 5.4).

Mutualism as described by Boucher *et al.* (1982) offers a means to frame a discussion of several types of interactions within the rhizosphere (Moore 1989; Wall and Moore 1999; Moore *et al.* 2003). Boucher *et al.* (1982) broadly defined mutualism as any set of interactions between organisms that benefited both, and in doing so identified two distinct forms of mutualism – symbiotic and non-symbiotic. Symbiotic mutualisms involve an intimate physical and physiological link between organisms, while non-symbiotic mutualisms involve benefits derived from means other than direct physical contact. Our chapter focused on symbiotic mutualisms and the non-symbiotic mutualisms of nutrient enrichment facilitated by the rhizosphere food web. Traditional theoretical treatments of symbiotic mutualism operating under the assumptions of local stability and equilibria concluded that the interactions are unstable when there are no constraints on growth in the face of disturbance (Pimm 1982). At first glance this conclusion appears at odds with the observation that mutualism is ubiquitous and for terrestrial ecosystems is a prominent feature of all major classifications of plant communities (Wall and Moore 1999). However, as our work indicates, mutualisms through their impact on nutrient cycling and availability place constraints on growth that can effect change in community structure and dynamics (Figure 5.6) when we move beyond models based on equilibria and local stability to ones that include non-equilibrium and transitions on persistent dynamics states (Hastings 1996).

Some years ago, in his tome entitled the *Strategy of Ecosystem Development*, Odum (1969) described stability in terms of nutrients cycling and the ability of a system to retain nutrients, based on empirical observations of the characteristics of ecosystems at later successional seres. May (1973) offered a decidedly different but contemporary approach to Odum's entitled *Stability and Complexity in Model Ecosystems*, that asked if community development and architecture were the result of a strategy or the result of a winnowing process of stable over unstable possibilities based on mathematical models of population dynamics. Subsequent studies have made connections between nutrients and mathematical attributes of models that are related indirectly or directly to stability (e.g., resilience and return times), interaction strengths, or eigenvalues of community and Jacobian matrices (DeAngelis 1975; Yodzis 1981; Moore *et al.* 1993; de Ruiter *et al.* 1995). Our empirical work and models suggest that the observed compartmentalized architecture is not only important to the transfer and retention of nutrients within the system, but to its stability as well. Our treatment also describes the rhizosphere food web as intricate and tightly coupled in a manner that facilitates the reciprocal transfer of limiting nutrients between producers and consumers that is decidedly based on mutualism and important to stability (Boucher *et al.* 1982; Coleman *et al.* 1983; Moore *et al.* 2003). We suggest that the changes in the fundamental architecture of the rhizosphere or alterations in the rates of nutrient transfer within that architecture alter key processes that drive ecosystem development and persistence (de Ruiter *et al.* 2005). Whether the result of a strategy (*sensu* Odum 1969) or a winnowing process of elimination (*sensu* May 1973), the connection between trophic structure, nutrient dynamics, and stability that we observe within the rhizosphere are more than coincidental.

ACKNOWLEDGEMENTS

This work was supported by grants from the National Science Foundation (DEB-9815925, DEB-0086599, and DEB-0120169), the National Center for Ecological Analysis and Synthesis (NCEAS), The US Bureau of Land Management, the Colorado Heritage Wildlife Foundation, and the European Science Foundation (ESF). Input and feedback for the chapter were provided by members of the NCEAS Detritus Dynamics and the ESF InterACT working groups. Special thanks to Zoe Cardon for organizing the symposium at August 2000: the annual meeting of the Ecological Society of America in Snow Bird, UT.

REFERENCES

Andrén, O., T. Lindberg, U. Bostrom, M. Clarholm, A.-C. Hansson, G. Johansson, J. Lagerlof, K. Paustian, J. Persson, R. Pettersson, J. Schnurer, B. Sohlenius, and M. Wivstad. 1990. Organic carbon and nitrogen flows. *Ecol Bull* 40:85–126.

Bardgett, R.D., D.A. Wardle, and G.W. Yeates. 1998. Linking above-ground and below-ground food webs. How plant responses to foliar herbivory influences soil organisms. *Soil Biology Biochemistry* 30:1867–1878.

Beare, M.H., D.C. Coleman, D.A. Crossley, P.F. Hendrix, and E.P. Odum. 1995. A hierarchical approach to evaluating the significance of soil biodiversity to biogeochemical cycling. *Plant Soil* 170:5–22.

Bender, E.A., T.J. Case, and M.E. Gilpin. 1984. Perturbation experiments in community ecology: Theory and practice. *Ecology* 65:1–13.

Boucher, D.H., S. James, and K.H. Keeler. 1982. The ecology of mutualism. *Annual Review of Ecology and Systematics* 13:315–347.

Bringhurst, R.M., Z.G. Cardon, and D.J. Gage. 2001. Galactosides in the rhizosphere: Utilization by *Sinorhizobium meliloti* and development of a biosensor. *Proceedings of the National Academy of Sciences of the USA* 98:4540–4545.

Brussaard, L., J.A. Van Veen, M.J. Kooistra, and J. Lebbink. 1988. The Dutch programme of soil ecology and arable farming systems. I. Objectives, approach and some preliminary results. *Ecological Bulletins* 39:35–40.

Brussaard, L., V.M. Behan-Pelletier, D.E. Bignell, V.K. Brown, W.A.M. Didden, P.J. Folgarait, C. Fragoso, D.W. Freckman, V.V.S.R. Gupta, T. Hattori, D.L. Hawksworth, C. Klopatek, P. Lavelle, D. Malloch, J. Rusek, B. Soderstrom, J.M. Tiedje, and R.A. Virginia. 1997. Biodiversity and ecosystem functioning in soil. *Ambio* 26:563–570.

Clarholm, M. 1985. Possible roles of roots, bacteria, protozoa, and fungi in supplying nitrogen to plants. In: Fitter, A.H., Atkinson, D., Read, D.J., and Usher, M.B. (eds) *Ecological Interactions in Soils: Plants, Microbes, and Animals.* Blackwell Scientific Publications, Oxford, UK, pp. 355–365.

Coleman, D.C. 1976. A review of root production processes and their influence on soil biota in terrestrial ecosystems. In: Anderson, J.M., and Macfadyen, A. (eds) *The Role of Terrestrial and Aquatic Organisms in Decomposition Processes.* Blackwell Scientific Publications, Oxford, pp. 417–434.

Coleman, D.C. 1996. Energetics of detritivory and microbivory in soil in theory and practice. In: Polis, G.A., and Winemiller, K.O. (eds) *Food Webs: Integration of Patterns and Dynamics.* Chapman Hall, New York, pp. 39–50.

Coleman, D.C., Reid, C.P.P., and C.V. Cole. 1983. Biological strategies of nutrient cycling in soil systems. In: MacFyaden, A., and Ford, E.D. (eds) *Advances in Ecological Research,* Vol 13. Academic Press, London, pp. 1–55.

Curl, E.A., and B. Truelove. 1986. *The Rhizosphere.* Springer, New York.

DeAngelis, D.L. 1975. Stability and connectance in food web models. *Ecology* 56:238–243.

DeAngelis, D.L. 1992. *Dynamics of Nutrient Cycling and Food webs.* Chapman and Hall, New York.

de Ruiter, P.C., J.C. Moore, K.B. Zwart, J.B. Bloem, L.A. Bouwman, J. Hassink, J.Y.C. Marinissen, W.A.M. Didden, J.A. De Vos, G. Lebbink, and L. Brussaard. 1993a. Simulation of nitrogen mineralization in the belowground food webs of two winter wheat fields. *Journal of Applied Ecology* 30:95–106.

de Ruiter, P.C., J.A. Van Veen, J.C. Moore, L. Brussaard, and H.W. Hunt. 1993b. Calculation of nitrogen mineralization in soil food webs. *Plant Soil* 157:263–273.

de Ruiter, P.C., A. Neutel, and J.C. Moore. 1994. Modelling food webs and nutrient cycling in agro-ecosystems. *TREE* 9:378–383.

de Ruiter, P.C., A. Neutel, and J.C. Moore. 1995. Energetics, patterns of interaction strengths, and stability in real ecosystems. *Science* 269:1257–1260.

de Ruiter, P.C., A. Neutel, and J.C. Moore. 1996. Energetics and stability in belowground food webs. In: Polis, G.A., and Winemiller, K.O. (eds) *Food Webs: Integration of Patterns and Dynamics.* Chapman Hall, New York, pp. 201–210.

de Ruiter, P.C., A. Neutel, and J.C. Moore. 1998. Biodiversity in soil ecosystems: The role of energy flow and community stability. *Applied Soil Ecology* 10:217–228.

de Ruiter, P.C., V. Wolters, J.C. Moore, and K.O. Winemiller. 2005. Food web ecology: Playing Jenga and beyond. *Science* 309:68–71.

Doles, J. 2000. A survey of soil biota in the arctic tundra and their role in mediating terrestrial nutrient cycling. *Masters Thesis,* Department of Biological Sciences, University of Northern Colorado, Greeley.

Farrar, J., M. Hawes, D. Jones, and S. Lindow. 2003. How roots control the flux of carbon to the rhizosphere. *Ecology* 84:827–837.

Foster, R.C. 1988. Microenvironments of soil microorganisms. *Biology and Fertility of Soils* 6:189–203.

Grayston, S.J., D. Vaughan, and D. Jones. 1996. Rhizosphere carbon flow in trees, in comparison with annual plants: The importance of root exudation and its impact on microbial activity and nutrient availability. *Applied Soil Ecology* 5:29–55.

Hastings, A. 1996. What equilibrium behavior of Lokta-Volterra models does not tell us about food webs. In: Polis, G.A., and Winemiller, K.O. (eds) *Food webs: Integration of Patterns and Dynamics*. Chapman Hall, New York, pp. 211–217.

Hastings, A., and Powell, T. 1991. Chaos in a three-species food chain. *Ecology* 72:896–903.

Hendrix, P.F., R.W. Parmelee, D.A. Crossley, D.C. Coleman, E.P. Odum, and P.M. Groffman. 1986. Detritus food webs in conventional and no-tillage agroecosystems. *Bioscience* 36:374–380.

Holling, C.S. 1959. The components of predation as revealed by a study of small mammal predation of the European pine sawfly. *Canadian Entomologist* 91:293–320.

Hunt, H.W., D.C. Coleman, E.R. Ingham, R.E. Ingham, E.T. Elliott, J.C. Moore, S.L. Rose, C.P.P. Reid, and C.R. Morley. 1987. The detrital food web in a shortgrass prairie. *Biology and Fertility of Soils* 3:57–68.

Ingham, R.E., J.A. Trofymow, E.R. Ingham, and D.C. Coleman. 1985. Interactions of bacteria, fungi, and their nematode grazers: Effects on nutrient cycling and plant growth. *Ecological Monographs* 55:119–140.

Jackson, R.B., J. Canadell, J.R. Ehleringer, H.A. Mooney, O.E. Sala, and E.D. Schulze. 1996. A global analysis of root distributions for terrestrial biomes. *Oecologia* 108:389–411.

Johnson, N., J.D. Hoeksema, L. Abbott, J. Bever, V.B. Chaudhary, C. Gehring, J. Klironomos, R. Koide, R.M. Miller, J.C. Moore, P. Moutoglis, M. Schwartz, S. Simard, W. Swenson, J. Umbanhowar, G. Wilson, and C. Zabinski. 2006. From Lilliput ot Brobdingnag: Extending models of mycorrhizal function across scales. *Bioscience* 56:889–900.

Lussenhop, J., and Fogel, R. 1991. Soil invertebrates are concentrated on roots. In: Keiser, D.L., and Cregan, P.B. (eds) *The Rhizosphere and Plant Growth*. Kluwer Academic Publishers, Dordrecht, p. 111.

May, R.M. 1972. Will a large complex system be stable? *Nature* 238:413–414.

May, R.M. 1973. *Stability and Complexity in Model Ecosystems*, 2nd edn, Princeton University Press, Princeton.

May, R.M. 1976. Simple mathematical models with very complicated dynamics. *Nature* 261:459–467.

McCann, K.S., A. Hastings, and G.R. Huxel. 1988. Weak trophic interactions and the balance of nature. *Nature* 395:794–798.

Moore, J.C. 1989. Influence of soil microarthropods on belowground symbiotic and non-symbiotic mutualisms. Interactions between soil inhabiting invertebrates and microorganisms in relation to plant growth. *Agricultural, Ecosystems and Environment* 24:147–159.

Moore, J.C., and H.W. Hunt. 1988. Resource compartmentation and the stability of real ecosystems. *Nature* 333:261–263.

Moore, J.C., and P.C. de Ruiter. 1991. Temporal and spatial heterogeneity of trophic interactions within belowground food webs. *Agricultural, Ecosystems and Environment* 34:371–394.

Moore, J.C., and P.C. de Ruiter. 1997. Compartmentalization of resource utilization within soil ecosystems. In: Gange, A., and Brown, V. (eds) *Multitrophic Interactions in Terrestrial Systems*. Blackwell Science, Oxford, pp. 375–393.

Moore, J.C., and P.C. de Ruiter. 2000. Invertebrates in detrital food webs along gradients of productivity. In: Coleman, D.C., and Hendrix, P.F. (eds) *Invertebrates as Webmaster in Ecosystems*, CABI, New York, pp. 161–184.

Moore, J.C., P.C. de Ruiter, and H.W. Hunt. 1993. Influence of productivity on the stability of real and model ecosystems. *Science* 261:906–908.

Moore, J.C., P.C. de Ruiter, H.W. Hunt, D.C. Coleman, and D.W. Freckman. 1996. Microcosms and soil ecology: Critical linkages between field studies and modelling food webs. *Ecology* 77:694–705.

Moore, J.C., D.E. Walter, and H.W. Hunt. 1988. Arthropod regulation of micro- and mesobiota in belowground food webs. *Annual Review of Entomology* 33:419–439.

Moore, J.C., K. McCann, H. Setälä, and P.C. de Ruiter. 2003. Top-down is bottom-up: Does predation in the rhizosphere regulate aboveground production. *Ecology* 84:846–857.

Moore, J.C., E. Berlow, D.C. Coleman, P.C. de Ruiter, Q. Dong, A. Hastings, N. Collins-Johnson, K.S. McCann, K. Melville, P.J. Morin, K. Nadelhoffer, A.D. Rosemond, D.M. Post, J.L. Sabo, K.M. Scow, M.J. Vanni, and D.H. Wall. 2004. Detritus, trophic dynamics, and biodiversity. *Ecology Letters* 7:584–600.

Moore, J.C., K. McCann, and P.C. de Ruiter. 2005. Modeling trophic pathways, nutrient cycling, and dynamic stability in soils. *Pedobiologia* 49:499–510.

Neutel, A.M., J.A.P. Heesterbeek, and P.C. de Ruiter. 2002. Stability in real food webs: Weak links in long loops. *Science* 296:1120–1123.

Odum, E.P. 1969. The strategy of ecosystem development. *Science* 164:262–279.

O'Neil, R.V. 1969. Indirect estimation of energy fluxes in animal food webs. *Journal of Theoretical Biology* 22:284–290.

O'Neill, R.V., D.L. DeAngelis, J.B. Waide, and T.F.H. Allen. 1986. *A Hierarchical Concept of the Ecosystem*, Princeton University Press, Princeton.

Paine, R.T. 1980. Food webs: Linkages, interaction strength and community infrastructure. *Journal of Animal Ecology* 49:667–685.

Parmelee, R.W., J.G. Ehrenfeld, and R.L. Tate. 1993. Effects of pine roots on microorganisms, fauna, and nitrogen availability in two soil horizons of a coniferous spodosol. *Biology and Fertility of Soils* 15:113–119.

Philips, D.A., H. Ferris, D.R. Cook, and D.R. Strong. 2003. Molecular control points in rhizosphere food webs. *Ecology* 84:816–826.

Pimm, S.L .1982. *Food Webs*. Chapman Hall, London.

Post, D.M., M.E. Conners, and D.S. Goldberg. 2000. Prey preference by a top predator and the stability of linked food chains. *Ecology* 81:8–14.

Reeves, F.B., D. Wagner, T. Moorman, and J. Kiel. 1979. The role of endomycorrhizae in revegetation practices in the semi-arid west. I. A comparison of incidence of mycorrhizae in severely disturbed versus natural environments. *American Journal of Botany* 66:6–13.

Reiners, W.A. 1986. Complementary models for ecosystems. *American Naturalist* 127:59–73.

Rosenzweig, M.L. 1971. The paradox of enrichment: Destabilization of exploitation ecosystems in ecological time. *Science* 171:385–387.

Schoener, T.W. 1974. Resource partitioning in ecological communities. *Science* 185:27–39.

Shaver, G.R., W.D. Billings, F.S. Chapin III, A.E. Giblin, K.J. Nadelhoffer, W.C. Oechel, and E.B. Rastetter. 1992. Global change and the carbon balance of arctic ecosystems. *Bioscience* 42:433–441.

Sterner, R.W., and J.J. Elser. 2001. *Ecological Stoichiometry*. Princeton University, Princeton.

Swartz, M.W., and J.D. Hoeksema. 1998. Specialization and resource trade: Biological markets as a model of mutualisms. *Ecology* 79:1029–1038.

Trofymow, J.A., and D.C. Coleman. 1982. The role of bacterivorous and fungivorous nematodes in cellulose and chitin decomposition in the context of a root [rhizosphere] soil conceptual model. In: Freckman, D.W. (ed.) *Nematodes in Soil Ecosystems*. University of Texas Press, Austin, pp. 117–137.

Van der Putten, W.H., L.E.M. Vet, J.A. Harvey, and F.L. Wäckers. 2001. Linking above- and belowground multitrophic interactions of plants, herbivores, pathogens, and their antagonists. *TREE* 16:547–554.

Vitousek, P.M., and L.R. Walker. 1989. Biological invasion by *Myrica faya* alters ecosystem development in Hawai'i: Plant demography, nitrogen fixation, ecosystem effects. *Ecological Monographs* 59:247–265.

Vitousek, P.M., L.R. Walker, L.D. Whiteaker, D. Mueller-Dombois, and P.A. Matson. 1987. Biological invasion of *Myrica faya* alters ecosystem development in Hawaii. *Science* 238:802–804.

Wall, D.W., and J.C. Moore. 1999. Interactions underground: Soil biodiversity, mutualism and ecosystem processes. *BioScience* 49:109–117.

Yodzis, P. 1981. The stability of real ecosystems. *Nature* 284:544–545.

Yodzis, P. 1988. Indeterminacy of ecological interactions, as perceived by perturbation experiments. *Ecology* 69:508–515.

Yodzis, P. 1996. Food webs and perturbation experiments: Theory and practice. In: Polis, G.A., and Winemiller, K.O. (eds) *Food Webs: Integration of Patterns and Dynamics*. Chapman Hall, New York, pp. 192–200.

CHAPTER 6

Understanding and Managing the Rhizosphere in Agroecosystems

Laurie E. Drinkwater and Sieglinde S. Snapp

6.1 INTRODUCTION

Agricultural systems represent the major form of land management, covering 5 billion hectares of the global terrestrial land area. The unintended consequences of agriculture extend well beyond agricultural landscapes and include environmental degradation and social displacement (Hambridge 1938; Friedland *et al.* 1991; Vitousek *et al.* 1997). Many have advocated the adoption of an ecosystem-based approach that would incorporate multifunctionality as an agricultural goal and entail broad application of fundamental ecological principles to food production (Dale *et al.* 2000; Drinkwater and Snapp 2007). This approach would aim to reduce external inputs and environmental degradation by increasing the capacity for internal, ecological processes to support crop production while contributing to other ecosystem services (Dale *et al.* 2000).

Most efforts devoted to managing the rhizosphere in agricultural systems have emphasized processes that contribute directly to maximizing yield within the context of resource-intensive cropping systems. Several excellent reviews are available covering the role of rhizosphere biology in promoting crop growth under the nutrient-rich conditions of high input agriculture (cf. Lynch 1990; Pinton *et al.* 2001). In particular, the biology of important root pathogens and plant-microbial N-fixing symbioses have been extensively studied within this context (Spaink *et al.* 1998; Whipps 2001). A smaller amount of rhizosphere research has focused on achieving modest improvements in yields under severe nutrient or water limitations that are commonly found in low-input, subsistence agroecosystems of the developing countries where farmers do not have access to purchased fertilizers and pesticides (Lynch 1990).

In this chapter, we will assess the current ecological understanding of the rhizosphere in agroecosystems and broaden the scope of rhizosphere contributions to encompass a variety of ecosystem functions beyond those directly related to maximizing crop growth and yields. Our aim is to examine the potential for rhizosphere processes and plant–microbial interactions to restore agroecosystem functions to reduce input dependency and environmental degradation. We begin with an inventory of how conventional, high-input management has altered the soil environment and biota in agroecosystems with particular emphasis on the consequences for the rhizosphere habitat. We then survey a range of rhizosphere processes and examine how current management practices enhance or hinder the process and evaluate the potential for improved functionality. Finally, we look ahead and discuss how management of the rhizosphere and plant–microbial interactions could be approached within multifunctional, ecologically sound agricultural systems of the future.

6.2   INTENSIVE AGRICULTURE: DELIBERATE AND INADVERTENT CONSEQUENCES FOR THE RHIZOSPHERE

The soil environment in agroecosystems reflects the legacy of the native ecosystem and past management combined with current management practices. In early farming systems, human modification was limited to altering plant species composition. As agriculture has evolved, the degree of intervention has grown steadily, culminating with the current, resource-intensive "Green Revolution" production systems where management interventions are often the dominant force shaping agroecosystem structure and function. Intentional management of the rhizosphere has focused mainly on biological control of root pathogens and enhancing obligate mutualisms and will be discussed later in this chapter. Here we briefly survey key modifications of the soil environment that result from a broad suite of management practices and their unintentional consequences for the rhizosphere habitat. In practice, farming systems fall along a continuum of intensity as depicted in Figure 6.1 and have varying impacts on rhizosphere processes. Our discussion will emphasize the situation in conventional, high-input annual systems since these production systems supply a substantial portion of food on a global basis, are continuing

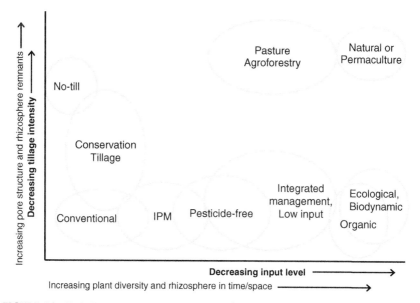

FIGURE 6.1   Variation in management intensity and consequences for the rhizosphere. Agricultural management systems are distinguished by the reliance on agrochemical inputs (fertilizers, pesticides, other agrochemicals) and tillage intensity. As the level of input dependance is reduced, the reliance on plant functional diversity generally increases as does the extent of the rhizosphere in space and time. Reduced dependence on tillage for annual crops (no-till) and perennial systems (pastures, agroforestry) are closest to unmanaged ecosystems in terms of soil physical environment and the presence of rhizosphere remnants as such as intact networks of root and hyphal pores (Williams and Weil 2004). Conventional production systems rely on high inputs and intensive tillage. Integrated pest management aims to substitute cultural practices and managed biodiversity for pesticides and has become a standard approach to managing insecticides efficiently in many cropping systems (Lewis *et al.* 1997). Pesticide-free agriculture is a grower-initiated approach largely oriented toward filling consumer demand for foods that are free of pesticides (Ott 1990). Integrated or low input systems combine agrochemical use with ecological or organic practices with the goal of reducing environmental impacts while achieving high yields (Reganold *et al.* 2001). Organic agriculture is the dominant from of ecologically based food production in the United States and seeks to minimize external inputs while avoiding all synthetic agrochemicals (National Organic Program 2005). Ecological and biodynamic systems originated in Europe and have an increased requirement for internalized N-fixation compared to US certified organic (International Federation of Organic Agricultural Movements 2005).

to expand in developing countries, and also have the greatest impact on the environment (Tilman 1999).

TILLAGE AND SOIL STRUCTURE

Use of tillage in agricultural production began with the development of the plow (~4000 BC in Mesopotamia and Egypt), which permitted large plots of soil to be intensively mixed and planted to a monoculture (Pryor 1985). Tillage remains a ubiquitous feature of nearly all-annual cropping systems. A variety of tillage technologies have been developed; however, most primary tillage involves mixing the top 15–25 cm of soil in preparation for planting. In addition to the periodic disruption of the soil environment, other consequences of tillage include radically altered pore volume and pore structure, reduced vertical stratification, destruction of biopores from past roots and hyphae, dispersal of microbial communities and fungal hyphae networks, and accelerated decomposition of soil organic matter (SOM; Buyanovsky and Kucera 1987; Douds *et al.* 1995; Baer *et al.* 2002). Following tillage, the soil tends to settle so that porosity is reduced compared to the original conditions of the native pre-tillage ecosystem. For example, the pore volume (percent of soil volume occupied by air or water) in native prairies of the Midwestern United States is approximately 60–70 percent compared to 45–50 percent in cultivated prairie soils (Baer *et al.* 2002). Thus, the reliance on tillage profoundly alters the soil environment in terms of atmospheric and water relations (drainage and water-holding capacity) while also disrupting processes that are influenced by legacy effects of former roots and fungal hyphae. No-tillage agricultural systems have been developed for annual crops such as the Midwestern grain systems of the United States. Even short periods without tillage extend the influence of root and hyphal remnants in soil (Douds *et al.* 1995). A recent study (Williams and Weil 2004) using a minirhizotron to monitor root growth discovered that root channels are recycled in annual rotations where the cash crop was planted without tillage following a *Brassica* cover crop (Figure 6.2). In practice, very few fields are maintained in continuous no-till beyond several years (c.f. Drinkwater and Snapp 2007) so only perennial agricultural systems such as pastures and some orchards maintain soil environments that approximate the native state in terms of the degree of physical disturbance and pore structure (see Figure 6.1).

NUTRIENT AVAILABILITY AND SOIL CHEMISTRY

Manipulation of soil chemistry began with the advent of liming to raise soil pH in early Roman agriculture and has grown to be a major component of soil fertility management in conventional agriculture so that soil pH is usually more neutral relative to native soils. In general, soil chemistry management aims to optimize the supply of nutrients to the crop. For the past 50 years, intensive agriculture has focused on supplying soluble, plant available forms of major nutrients combined with manipulation of soil pH and additions of micronutrients as indicated by soil tests (Drinkwater and Snapp 2007). In wealthier, industrialized countries major nutrients are generally supplied in surplus quantities as soluble, inorganic fertilizers resulting in cropping

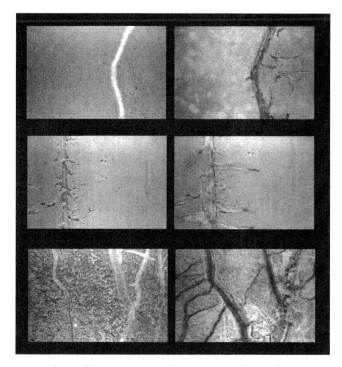

FIGURE 6.2 Paired minirhizotron images showing roots of a canola cover crop (left) in compacted plowpan soil (spring) and soybean roots (right) observed at the same locations in the soil a few months later. The roots can be seen to follow channels made by the preceding canola roots (after Williams, S.M. and Weil, R.R. (2004) "Crop cover root channels may alleviate soil compaction effects on soybean crop." *Soil Science Society of America Journal*; with permission).

systems which are maintained in a state of nutrient saturation, particularly when the cash crop is present. Compared to unmanaged terrestrial systems, the concentrations of soluble, inorganic forms of major nutrients such as N and P in these agricultural systems are often several orders of magnitude greater. In contrast, agroecosystems of poorer developing countries do not have access to manufactured fertilizers and are often producing crops in soils that have depleted nutrient pools from long histories of farming without adequate nutrient return to fields.

CARBON FLOW AND SOIL ORGANIC MATTER

The use of tillage has major consequences for C distribution and turnover. Tillage eliminates the O horizon and accelerates litter decomposition rates by mixing newly introduced litter with soil (Buyanovsky and Kucera 1987).

Furthermore, the advent of fertilizers and herbicides made it unnecessary to grow cover crops and forages in rotation with cash crops and permitted the widespread adoption of the simplified crop sequences that are prevalent today (Drinkwater and Snapp 2007). These rotations typically include bare fallow periods (when land is maintained without any growing plants) in between cash crops. As a result, the time frame of actively growing plants in annual agriculture is commonly limited to 4–6 months per year, decreasing the rhizosphere habitat, C-fixation, and inputs of labile C in space and time. Tillage, combined with soluble N additions and the relatively labile composition of crop residues returned to the soil, fosters rapid turnover of particulate organic soil C pools (Buyanovsky and Kucera 1987) while decomposition of the humified fraction may decrease (Neff et al. 2002), shifting the distribution of C pools so that labile, particulate OM is proportionately reduced compared to humified OM (Wander 2004). The reduction in total SOM combined with the disproportionate impact on labile C pools increases the severity of C-limitation in bulk soil and exacerbates the tendency for nutrient saturation while also modifying microbial habitat distribution.

CONSEQUENCES FOR THE SOIL BIOTA AND THE RHIZOSPHERE COMMUNITY

The abiotic changes outlined above restructure the distribution and frequency of microbial soil habitats (i.e., rhizosphere, aggregates, particulate organic matter, and biopores) in agroecosystems and lead to shifts in species abundance and richness of the soil biota. Management interventions such as tillage, crop species composition, and soil amendments act in concert with the background soil environment to alter the indigenous biota in bulk soil (Guemouri-Athmani et al. 2000), and to influence rhizosphere community composition (c.f. Buckley and Schmidt 2003; da Silva et al. 2003; Buenemann et al. 2004). Studies comparing the resident microbial community composition across managed ecosystems suggest that soil type is the most important factor, followed by land use and management history (Salles et al. 2004; Bossio et al. 2005). In the short term, the influence of plants is most significant for the plant-associated habitats such as the rhizosphere and rhizoplane (Salles et al. 2004); however, as the longevity of crop cultivation increases, plant species impacts on microbial community become more detectable (Guemouri-Athmani et al. 2000; Schloter et al. 2000). The effects of cultivation on microbial community structure in bulk soil appear to be long-lasting and can still be detected years after agricultural management has ended (Buckley and Schmidt 2003).

It is clear that under intensified agriculture, the rhizosphere community is faced with a unique soil environment that differs substantially from the one in which plant–microbial interactions originally evolved. Ecosystem services that were once supplied by plants and associated soil organisms are now largely provided through a variety of inputs. Biotic functional diversity has been replaced by increased intervention including tillage, and the use of soluble fertilizers and pesticides (Drinkwater and Snapp 2007). In essence, it is the management system that has created the dependence upon many of the agricultural inputs that target the belowground system in agriculture, similar to the pesticide treadmill that was first proposed in the 1960s to describe the increased dependence on insecticides created by chemical control of aboveground herbivorous arthropods (Smith and van der Bosch 1967). The use of tillage necessitates the need for continued tillage due to diminished SOM and degraded soil structure (Topp *et al.* 1995). The high concentrations of plant-available nutrients in space and time may reduce the role of mutualist rhizosphere organisms since energetics favor plant acquisition of these soluble nutrients which are supplied in quantities surpassing crop needs (see Chapter 4). Finally, alterations in the soil environment combined with simplified rotations often increase the frequency and severity of pathogen infections leading to dependence on broad-spectrum fungicides such as methyl bromide (Cook 1993; Abawi and Widmer 2000).

6.3   RHIZOSPHERE PROCESSES AND AGROECOSYSTEM FUNCTION

It is against this backdrop of a highly modified soil environment and the cascading effects on soil biota that we examine the rhizosphere in agriculture and consider how to redirect management to restore rhizosphere processes and agroecosystem function. Rhizosphere microorganisms and their associated primary producers contribute both directly and indirectly to a wide range of ecosystem functions (Figure 6.3). Processes such as aggregation, nutrient cycling, hydrology, and C-storage are jointly mediated by plants and soil organisms through interactions in the rhizosphere, although the significance of rhizosphere contributions has generally been diminished in agroecosystems by inputs and other interventions (Figure 6.1).

RHIZOSPHERE-MEDIATED AGGREGATION

Soil aggregation determines the pore structure and dispersion resistance of soil and is a fundamental driver of soil and ecosystem functioning. The proportion of soil particles sequestered in aggregates contributes to the movement and storage of water, soil aeration, and species composition and distribution of soil organisms. These factors interact with one another and influence

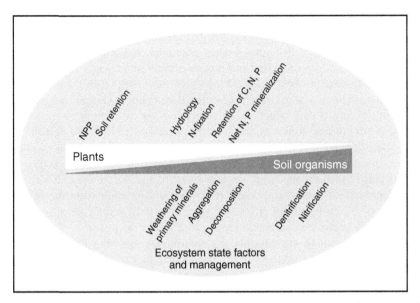

FIGURE 6.3 Rhizosphere processes contribute to a variety of ecosystem functions and services and are the outcome of plant–microbial collaborations. Some, such as soil retention, are strongly governed by plant species and others such as nitrification are largely controlled by microorganisms.

numerous ecosystem processes including: (1) water-holding capacity, infiltration and erosion; (2) temporal and spatial distribution of anaerobic conditions; (3) growth of plant roots and fungal hyphae; and (4) the cycling and storage of nutrients and carbon (Angers and Caron 1998). The process of aggregate formation is particularly important in agroecosystems where tillage periodically disrupts soil structure and accelerates the breakdown of macro-aggregates and physically protected labile SOM. While parent material and clay surface chemistry regulate micro-aggregate formation, plants and soil organisms are the major drivers of macro-aggregate formation (Tisdall and Oades 1982). As a result, the feedbacks between plants, soil organisms and soil structure are key regulators of productivity and biogeochemical functioning of agroecosystems. In soils with a high proportion of fine silt and clay particles, aggregation is a prerequisite for life because most plants require some level of aggregate structure to create pores for aeration and water flow in order to grow in these soils.

The potential for agricultural plants and their associated rhizosphere organisms to foster aggregate formation has been studied since the early 1980s (Reid and Goss 1981; Tisdall and Oades 1982) and is recently reviewed by Angers and Caron (1998). Most of this research has focused on the effect of different plant species on water-stable aggregation in bulk soil which reflects root litter and rhizosphere processes. In general, perennial forage crops and annual legumes and grasses tend to increase water-stable aggregation compared to cash crops or bare fallow (Reid and Goss 1981; Angers and Caron 1998). These are the agricultural plants that have been removed from most rotations in favor of simplified rotations enabled by modern agriculture. Maize, tomato, and wheat actually decreased aggregate stability, while growth of perennial ryegrass and alfalfa tended to increase it. Rhizosphere-mediated impacts on aggregate stability in bulk soil accrue over time and have been related to plant parameters such as total root biomass or root length (Rillig *et al.* 2002), microbial polysaccharides produced in the rhizosphere (Reid and Goss 1981), and more recently, fungal populations associated with the rhizosphere (Haynes and Beare 1997; Rillig *et al.* 2002) or fungal products such as glomalin (Wright *et al.* 1996).

The role of plant–microbe symbiosis in aggregate formation is most extensively documented for arbuscular mycorrhizal fungi (AMF) which are an important biotic regulator of water-stable aggregation in bulk soil. The discovery of glomalin, a collection of iron-containing glycoproteins produced by AMF (Wright *et al.* 1996), has provided an unprecedented opportunity to study the ecology of biotic aggregate formation. To date, glomalin has been found in virtually all soils tested for the glycoprotein although the quantity can vary from up to $100\,\mathrm{mg\,g\;soil}^{-1}$ in tropical forest soils to $3-4\,\mathrm{mg\,g\;soil}^{-1}$ in temperate agricultural soils (Wright and Anderson 2000; Rillig *et al.* 2001). Glomalin is a moderately stable component of the SOM, with a mean turnover time reported to range from 6 to 40 years. The structure of glomalin has not been characterized so the true function of the glycoprotein remains unresolved and is an active area of research.

More recently, the role of rhizosphere bacteria in promoting aggregation of soil within close proximity to roots, that is rhizosphere or root-adhering soil, has been investigated. Several rhizosphere organisms that foster aggregate formation through production of exopolysaccharides (EPS) have been identified and linked to improved soil structure of root-associated soil (Gouzou *et al.* 1993; Alami *et al.* 2000). The EPS-producing bacterium *Paeni bacillus polymyxa* (strain CF43), an N-fixing bacterium endemic to the wheat (*Triticum aestivum* L.) rhizosphere, fosters significant increases in the aggregation and water-holding capacity of soil adjacent to roots (Gouzou *et al.* 1993). In a study of sunflower (*Helianthus annuus* L.) and an EPS-producing *Rhizobium* sp. (Strain YAS34) isolated from the sunflower rhizosphere, inoculation with this organism resulted in increased abundance in the rhizosphere, modified soil structure and water-holding capacity around the root system, and a corresponding improvement in the drought resistance of the plant (Alami *et al.* 2000). This type of localized modification of soil structure through the production of EPS appears to be important for non-irrigated agricultural systems at the scale of individual plants. It is likely to be most significant for short-term modifications of water movement and storage since these polysaccharides are readily decomposed and probably have a much shorter mean residence time than glomalin-type substances.

Questions about the function of these compounds, their evolutionary significance, and the mechanisms that control their production remain unanswered. This information coupled with a greater understanding of the extent to which cultivar selection, tillage, and other interventions may have inadvertently influenced crop–microbial interactions that foster aggregate formation is crucial for intentional management of biotic aggregation. The promotion of aggregation through production of complex, extracellular polymers has often been viewed as a secondary consequence of release of these substances in the environment. Clearly, bacterial production of EPS is a widespread phenomenon occurring across microbial habitats associated with the formation of biofilms (Morris and Monier 2003). However, given the prevalence of organisms that are able to release copious amounts of these compounds into the soil and the benefits that accrue through improvements in soil structure, it is possible that under certain conditions soil structure modification is the major function of these compounds. Although a systematic study of the abundance of EPS-producing species across ecosystems that vary in terms of clay content has not been conducted, some evidence suggests that at least within bacterial species, strains that are present in high-clay soils tend to produce significant amounts of EPS (Achouak *et al.* 1999) and promote aggregate formation (Gouzou *et al.* 1993). One in vitro experiment demonstrated that glomalin production by *Glomus intraradices* is extremely plastic and appears to respond to environmental conditions such as pore structure (Rillig and Steinberg 2002). In this study, the production of glomalin was increased under unfavorable growing conditions simulating a soil structure lacking sufficient pores. These findings are intriguing; however, more research is needed to conclusively determine the primary function of these extracellular compounds.

DECOMPOSITION AND NET MINERALIZATION OF NUTRIENTS

Plant-mediated decomposition and corresponding mineralization of nutrients via the rhizosphere ("microbial loop" or "priming effect", see Chapters 2 and 3) is not considered to be important in conventional agriculture and hence, deliberate management of this process has not been attempted. While some plants are able to produce and secrete enzymes required for P mineralization (Vance *et al.* 2003), release of nutrients from organic compounds is largely carried out by heterotrophic microorganisms through the production of extracellular enzymes that can attack polymers and release small, soluble molecules. The role of plant–microbial interactions in accessing organic nutrient pools is considered to be of central importance in organically managed systems (Drinkwater 2004). Although SOM pools are not generally the target of conventional soil fertility management, it is clear that the microbial loop may serve as a significant source of nutrients, particularly N, even in cropping systems receiving large fertilizer additions. Despite application of luxurious amounts of N and use of refined best management practices, crops still acquire 40–80 percent of their N from endogenous soil reserves, and an average of 50 percent of the N applied is lost from agricultural landscapes (Tilman 1999).

Clearly, greater reliance on plant-mediated mineralization for nutrient acquisition in agroecosystems would reduce the potential for nutrient losses due to the tight coupling between the release of soluble, potentially mobile nutrient forms and plant uptake in the rhizosphere. This could be particularly advantageous in the case of N, which is highly susceptible to loss once it is converted to inorganic forms. Inorganic N pools can be extremely small while high rates of net primary productivity (NPP) are maintained if N-mineralization and plant assimilation are spatially and temporally connected in this manner.

We expect that during millions of years of coevolution plant–microbial feedbacks have evolved to regulate this co-dependency. Until the advent of Haber-Bosch N, the presence of inorganic N was indicative of net mineralization (with the exception of soils where NH_4^+ is present in clays). Thus, plants could increase their access to N through root proliferation and exudation of labile C to support decomposition when inorganic N patches were encountered. In split root studies, rhizodeposition is increased by roots exposed to greater concentrations of inorganic N compared to roots from the same plant that are under low inorganic N conditions (Paterson 2003). We believe it is safe to say that the feedback mechanisms regulating this exchange are not fully understood and that the coupling of plant–microbial nutrient flows has probably been unintentionally modified in high-input agricultural systems. Furthermore, it is unclear how crop breeding, which has occurred primarily under nutrient saturated conditions, has affected the ability of crops to access these pools of organic N via this mechanism. While crops certainly stimulate microbial decomposition of SOM (Clarholm 1985; Paterson 2003), selection in soil environments where inorganic N and P are supplied in surplus quantities would tend to favor the development of cultivars that did not squander fixed C to obtain N or P. In fact, it is possible that the combination of a modified soil environment and crop selection for such an environment has undermined the capacity of some crops to access certain organic nutrient reserves.

While we can speculate about the mechanisms regulating plant–microbial interactions that influence decomposition, many questions remain to be answered if we are to effectively manage this process in agroecosystems. In particular, it will be important to identify the SOM pools accessed by plant-mediated decomposition in order to manage agroecosystems to enhance these reservoirs without fostering increases in net N mineralization in the absence of plants. Our current understanding of decomposition energetics suggests that labile C is needed in order to support the breakdown of the large reservoirs of humus which consists of complex polymers with a C:N ratio of about 12:1. The importance of rhizosphere habitats in fostering decomposition of chemically recalcitrant substrates is supported by bioremediation studies which have demonstrated that decomposition of soil contaminants such as polycyclic aromatic hydrocarbons is accelerated in the rhizosphere (Siciliano *et al.* 2003). In addition to targeted management of SOM pools, other attributes such as food-web structure could also be influenced by management to optimize this process.

RHIZOBIAL AND MYCORRHIZAL ASSOCIATIONS

The specificity of the legume and rhizobia association has been exploited by farmers and agricultural scientists for centuries. Application of Rhizobia inoculum to the seeds of leguminous species is the most widely practiced, conventional agricultural technology used to deliberately manipulate rhizosphere microorganisms. This direct biological intervention has been credited with enhancing N-fixation from 30 to 75 percent in grain legumes (Moawad *et al.* 1998). However, indigenous strains of Rhizobia are often more effective at colonizing nodules than inoculated strains, even if the seed is inundated with Rhizobia inoculum. The interaction of focal plant with the bacterial inoculum, and the outcome in terms of colonization and development of a symbiotic organ such as nodules, are highly dependent on space and time. For instance, the community of nodule inhabitants is significantly influenced by rhizosystem architecture in inoculated soybeans (Espinosa-Victoria *et al.* 2000). Nodules located near the central root system are developed through plant symbiotic interactions with inoculated *Rhizobium* sp., while external nodules far from the central axis are likely to be inhabited by indigenous, and often ineffective, *Rhizobium*.

Indigenous rhizosphere populations generally resist invasion by inoculated organisms in the absence of host-microorganism specificity. This is illustrated by the widespread failure of efforts to manage arbuscular mycorrhizae in agricultural systems through inoculation-based technologies (Hamel 1996). There are exceptions, usually involving inundation of young, uncolonized tissues in an environment with few established organisms. Examples include inoculation of seeds or mycorrhizal treatment of horticultural plantings at mine rehabilitation sites, containerized systems or seriously degraded and fumigated soils (Jeffries *et al.* 2003). With the notable exception of the legume–Rhizobia association, inoculation techniques have not led to consistent or persistent effects on nutrient availability in conventional agriculture. A promising area of research is to examine the potential to manage these mutualisms in low-input and organic systems that provide an energetically and biologically favorable environment for displacing or augmenting indigenous micro-flora and fauna, compared to conventional agriculture (Kumar *et al.* 2001).

Agricultural management practices have profound indirect consequences on these rhizosphere mutualisms. Reliance on soluble nutrients markedly alters community dynamics in the rhizosphere and may have inadvertently selected for ineffective mycorrhizal and legume–*Rhizobium* symbioses in modern agricultural systems. A case can be made that the evolution of plant–microsymbiont relationships has been mediated by agricultural practices, many of which favor parasitism over mutualism (e.g., Kiers *et al.* 2002). Mycorrhizal species composition in high nutrient input corn systems has been shown to favor ineffective strains (Douds *et al.* 1995), but there has been very limited research on the mutualist to parasitic role of microbial associations in agroecosystems.

The N-balance in agricultural systems is profoundly influenced by the regulation of the N_2-fixation process by soluble nitrogen. There is genetic variation in both plant host and *Rhizobium* bacteria for tolerance of the N_2-fixation process to the presence of nitrate, but in the vast majority of cases the presence of nitrate is highly suppressive to symbiotic N_2-fixation (c.f. Kiers *et al.* 2002). Indirect consequences of the suppression of N_2-fixation by soil N may include suppression of mycorrhizal function since flavonoids that induce nodulation also stimulate hyphal growth of the AMF (Rengel 2002).

Nutrient input level is a major regulator of plant–mycorrhizal symbiosis (see Chapter 4). Application of inorganic P has been widely shown to directly suppress mycorrhizal infection of roots (Jasper *et al.* 1979), and to suppress function of the plant–mycorrhizal symbiosis in maize and soybean (McGonigle *et al.* 1999). Disturbance from tillage is another factor that reduces the presence of mycorrhizal symbiosis (Galvez *et al.* 2001). Further study of the intermediate and longer-term consequences of agricultural management practices on plant symbioses is urgently required.

BIOLOGICAL CONTROL AND COMMUNITY ECOLOGY OF THE RHIZOSPHERE

In general, conventional agricultural practices stimulate facultative saprophytic pathogens and increase crop susceptibility to disease. The edaphic environment is modified as shown in Figure 6.1, with high inorganic nutrient availability and low diversity carbon inputs associated with conventional agricultural systems. This profoundly influences substrate, habitat availability and microbial community dynamics (Hoitink and Boehm 1999). These environmental modifications in conjunction with short rotations are the root of many soilborne disease problems. This is evident in intensively managed, high-value vegetable crops where reliance on fumigation, multiple tillage operations and high rates of fertilizer is often associated with compacted soils, low levels of soil microbial activity, and recurring root health problems (Abawi and Widmer 2000).

The use of inoculation with beneficial, biological control organisms that will colonize the rhizosphere shows some promise as a means to suppress plant disease (Cook *et al*. 1993). Successful application has been rare, although a notable exception is inoculation of plant habitats with limited colonization such as seeds and emerging radicle. For example, application of *Pseudomonas fluorescens* to tomato seeds reduced development of the pathogen *Pythium ultimum*. Suppressive capacity was linked to siderophore production and established presence of *P. fluorescens* (Hultberg *et al*. 2000).

Efforts to reduce soilborne diseases by modifying the soil environment have also met with some success. Management that augments the time frame of living cover and diversity of carbon inputs is associated with enhanced activity and presence of soil microorganisms. If non-pathogenic rhizosphere organisms are well established, this will tend to suppress soilborne disease organisms through mechanisms such as competition for resources and habitat (Whipps 2001), antagonistic compounds (Robleto *et al*. 1998), degradation of pathogenicity factors or pathogen cell walls, promotion of vigorous, healthy roots (Snapp *et al*. 1991, 2003), and induction of systemic resistance in the target plant against the pathogen (van Wees *et al*. 1999). Soilborne phytopathogens encounter antagonism from rhizosphere microorganisms before, during, and after primary infection and secondary spread within the root. Readers are refereed to recent reviews which focus on the mechanisms of suppressive soils (e.g., Sturz and Christie 2003).

A well-studied example of rhizosphere occupants and consequences for soilborne disease is take-all (*Gaeumannomyces graminis* var. *tritici*) in wheat, one of the most important, devastating fungal diseases in cereal production around the world (Cook 1993). Wheat is a rare case where temporal monoculture has proven beneficial to pathogen suppression. Generally, after initial severe outbreaks, the disease is suppressed in continuous wheat monocultures through a phenomenon known as take-all decline. Altered population dynamics of rhizosphere bacteria are consistently associated with suppression of take-all, and most recently, Pseudomonads that secrete 2,4-diacetylphloroglucinol (2,4-DAPG), a compound that directly inhibits *G. graminis* var. *tritici* have been identified as the major antagonists (Gardener and Weller 2001; Mazzola 2004). As the longevity of a continuous wheat monoculture increases, 2,4-DAPG producing Pseudomonads become more abundant (Gardner *et al*.

2001). The control of *G. graminis* in wheat by 2,4-DAPG strains has been documented in disparate geographical regions and can be enhanced by management practices (i.e., no-till enhances the development of suppressive populations of Pseudomonads; Mazzola 2004). Finally, wheat cultivars play a role in determining the particular Pseudomonad strains that will become most abundant (Mazzola *et al.* 2004). This case can serve as a model example of the potential for rhizosphere ecology to be applied in managing complex plant–microbial interactions so that chemical controls are not needed.

PLANT SPECIES AND CULTIVAR EFFECTS ON RHIZOSPHERE PROCESSES

It is well known that plants exert an influential role on rhizosphere community composition (Chapter 1), and that the selection for particular microbial assemblages in the rhizosphere eventually impacts the microbial community structure in bulk soil (Schloter *et al.* 2000). This is important in the context of agricultural systems because it suggests that crop rotations can be intentionally designed to manage the resident microbial and rhizosphere communities. Indeed, this strategy has been employed in agriculture in a rudimentary fashion though the use of rotation to reduce the severity of soilborne diseases even before the identity of the pathogens was known (Cook 1993). Hawkes *et al.* (Chapter 1) have reviewed the literature on the role of plant species in determining rhizosphere microbial community composition. Here we examine plant influences on rhizosphere communities in an agricultural context with particular emphasis on the consequences of crop breeding and the role of plant intraspecific genotypic variation in regulating rhizosphere microbial community composition.

Plant selection in the last half-century has occurred almost entirely under management regimes that include fumigated soils with luxurious additions of nutrients and sufficient water (Boyer 1982). This strategy of reducing environmental variation by providing ample resources reduces gene by environment interaction, and enhances the power of selection for specific traits (Boyer 1982). Growing evidence suggests that this approach has altered belowground function in crops and may have selected against traits that allow crops to maintain productivity in environments that differ from those created by high-input systems (Jackson and Koch 1997; Bertholdsson 2004). While this is an interesting hypothesis, it has yet to be proven. In many cases, the underlying phenotypic alterations that have contributed to improving yield potential under these high-input conditions have not generally been identified (Boyer 1982); however, there are a few examples where modification of belowground plant traits of modern cultivars has been demonstrated (Jackson 1995; Briones *et al.* 2002;

Bertholdsson 2004). In lettuce, plant-breeding approaches have altered root architecture and the ability of root branching to respond to soil environmental conditions such as nutrient limitation (Jackson 1995; Jackson and Koch 1997). A study comparing 137 barley cultivars representing a time span of 100 years of crop selections found that allelopathic abilities resulting from root exudates are reduced in modern hybrids compared to older varieties (Bertholdsson 2004). In this study, alleopathic activity of roots decreased by 32–85 percent in modern hybrids compared to that of the older cultivars. Finally, in wheat and barley, increased plasticity in root hair length has been linked to improved capacity to acquire soluble P (Gahoonia *et al.* 1999).

Studies comparing the role of intraspecific genetic variation on rhizosphere microbial community composition have reported varying results. Some report little or no detectable cultivar effect (c.f. Devare *et al.* 2004) while others find substantial differences in microbial community composition in the various plant-associated habitats (Germida and Siciliano 2001; Briones *et al.* 2002; da Silva *et al.* 2003; Mazzola *et al.* 2004). Many of these studies do not include information about the degree of genotypic variation among the cultivars studied making it difficult to interpret contradictory results. Since the development of genetically modified crops, there have been numerous studies investigating the effect of these new genotypes on plant-associated soil microbes. Studies comparing the original non-GMO hybrid to the GMO version in terms of root-associated microbes report variable results; however, single gene modifications that are expressed in belowground functions frequently do result in detectable modifications of rhizosphere community composition (cf. Dunfield and Germida 2004). Comparisons across cultivars with noticeable phenotypic variability in systemic growth characteristics have detected significant differences in rhizosphere community composition (da Mota *et al.* 2002; da Silva *et al.* 2003; Mazzola *et al.* 2004). In a study comparing Pseudomonad rhizosphere populations in five wheat cultivars, Mazzola *et al.* (2004) reported that the cultivars differed in their capacity to select for populations of 2,4 DAPG producing *Pseudomonas* sp. Comparisons of traditional or pre-industrial cultivars to modern, highly selected varieties frequently detect differences in rhizosphere community composition (Germida and Siciliano 2001; Briones *et al.* 2002, 2003). As more sophisticated molecular techniques become widely available, a picture is emerging which suggests that more closely related cultivars will have greater similarities in rhizosphere community composition compared to more distantly related cultivars (Schloter *et al.* 2000; da Mota *et al.* 2002; da Silva *et al.* 2003).

Although it is becoming well accepted that intraspecific differences in plant genotypes influence rhizosphere community composition, the consequences for rhizosphere function are rarely understood. An excellent example linking plant genotypic and phenotypic differences to rhizosphere processes occurs in ammonia-oxidizing bacteria (AOB) populations in the rhizosphere of traditional versus modern rice cultivars (Briones *et al.* 2002, 2003). Efficient management of N in rice paddies is particularly challenging because rice paddies are typically maintained under flooded conditions and are essentially anoxic below the soil–water interface and AOB abundance is significantly enhanced in the rice rhizosphere (Briones *et al.* 2002). Using fluorescence in situ hybridization (FISH) with rRNA-targeted probes specific for the AOB to characterize AOB populations in the rhizosphere and on the rice root surface, Briones *et al.* (2002) reported differences in the total abundance and species composition of the rhizoplane microbial communities across cultivars (Table 6.1). The greater abundance of the faster-growing *Nitrosomonas* spp. can be partially attributed to the increased secretion of O_2 by cv. IR63087-1-17 compared to cv. Mahsuri (Briones *et al.* 2002). Shifts in the dominant AOB population were accompanied by greater abundance of heterotrophic bacteria in the rhizoplane of cv. Mashuri suggesting that other factors such as differences in root exudates and N dynamics (i.e., NH_4^+ assimilation by the plants and heterotrophs in the rhizosphere) may also contribute to the observed distributions of AOB species and nitrification rates (Briones *et al.* 2003). This case establishes the potential for deliberate management of biogeochemical processes through cultivar–microbe interactions while also illustrating the complexities involved in manipulating rhizosphere function.

TABLE 6.1 Comparison of rice cultivars, their rhizosphere composition, and biogeochemical function (from Briones *et al.* 2002, 2003)

	Improved traditional	Modern hybrid
Cultivar	Mahsuri	IR63087-1-17
Fertilizer use efficiency	Able to use either NH_4 or NO_3	Greater efficiency with NH_4 application
Rhizosphere environment	Roots are less permeable to O_2	Roots leak more O_2
Rhizosphere community composition[1]	Rhizoplane is dominated by heterotrophs	Heterotrophs may be less abundant compared to Mahsuri rhizoplane
Most abundant ammonia oxidizing bacteria	*Nitrosopira* sp. (able to grow at lower substrate concentrations, K strategist)	*Nitrosomonas* sp. (fast growing, R strategist)
Nitrogen cycling[2]	Nitrification is not detectable	Nitrification rate: $1.2\,\mu g\,N\,g\,soil\,day^{-1}$

[1] Detection and characterization of ammonia-oxidizing bacteria by PCR-DGGE targeting the *amoA* gene. Quantification of AOB in rhizosphere and rhizoplane was based on FISH.
[2] Based on in situ field experiments using the ^{15}N pool dilution method.

Finally, another area of interest from the standpoint of agriculture addresses the question of whether or not there are non-obligate, root-associated microorganisms that are endemic to the rhizosphere of particular plant species and the extent to which crops can modify these populations. Resolving this question could contribute to the development of breeding strategies that target crops and their associated microorganisms. The evidence that both interspecific and intraspecific composition of rhizophere associated assemblages are tightly coupled to the host plant genotype is increasing (Schloter *et al.* 2000; Mazzola *et al.* 2004). Wheat offers a particularly interesting system for consideration of this question since it is one of the oldest agricultural plants and is currently cultivated across a wide variety of climates and soil types. A number of microorganisms have consistently been found in rhizospheres of wheat growing in geographically dispersed soils from diverse environments (Schloter *et al.* 2000; Mazzola 2004). *P. polymyxa* is a free-living N-fixer that has been found in the wheat rhizosphere in North America (Nelson *et al.* 1976), Europe (Heulin *et al.* 1994), and Africa (Guemouri-Athmani *et al.* 2000), and the rhizosphere populations of this organism appear to be adapted to the wheat rhizosphere (Guemouri-Athmani *et al.* 2000). One study conducted in Algerian soils that had been under wheat cultivation from 5 to 2000 years, showed that the genetic composition of *P. polymyxa* populations varied across the chronosequence (Guemouri-Athmani *et al.* 2000). The longevity of wheat cultivation was correlated with decreased phenotypic and genetic diversity and higher frequency of N-fixing strains of *P. polymyxa* in rhizosphere-associated soil. The rhizosphere assemblage consisting of the wheat pathogen *G. graminis* var. *tritici* and the populations of 2,4 DAPH producing *Pseudomonas* spp. antagonists is also found in the wheat rhizosphere in widely distributed geographic locations (Mazzola 2004). These examples, while rather limited, demonstrate similarities in rhizosphere assemblages when plants of the same species are grown in a broad range of environments and supports the idea that plant genotype is the major driver influencing the formation of unique rhizosphere assemblages. The coupling between both inter- and intraspecific genetic diversity of plants and their associated rhizosphere organisms provides support for the extended phenotype concept (Dawkins 1982). This idea that a single, dominant species can influence community-scale evolutionary processes may be important in some ecosystems has been supported by recent studies of intraspecific genetic variability in plants and their associated aboveground arthropods (Whitham *et al.* 2003; Johnson and Agrawal 2005) and may also be a useful concept to apply to agricultural systems where monocultures are the norm. Further understanding of evolutionary mechanisms governing rhizosphere populations and plant–microbial interactions will be crucial in the development of crop-breeding strategies that target processes occurring in the rhizosphere.

6.4 THE FUTURE OF THE RHIZOSPHERE IN ECOLOGICAL AGRICULTURE

The management of biocomplexity to promote ecological processes and restoration of agroecosystem function will require the development and application of ecosystem-based management strategies. Ecosystem management is a land-management approach that (1) takes into account the full suite of organisms and ecosystem processes; (2) applies the concept that ecosystem function depends on ecosystem structure and diversity; (3) recognizes that ecosystems are spatially and temporally dynamic; and (4) includes sustainability as a primary goal (Dale *et al.* 2000). Application of this approach will require that we redirect management practices to create a soil environment that is more conducive to supporting potential contributions from rhizosphere processes. Opportunities to redirect management include changes in the way we approach cultivar selection, repopulating annual systems with plants in space and time and integration of rotation, tillage and soil-amendment practices.

SELECTION FOR PLANT–MICROBE CONSORTIA

We see many opportunities for plant breeding to enhance plant–microbial interactions in ways that contribute to restored ecosystem functions. First, the traditional breeding framework views plants as single organisms, and as a result, selection of crops and mutualists is often conducted separately. This book supports the notion that plants are more accurately viewed as a consortium consisting of a primary producer and many species of associated microbes (or depending on your bias, microorganisms and their associated plants!). Agricultural breeding programs should select for well-adapted consortia that can achieve necessary levels of primary productivity while maintaining ecosystem services through optimization of plant microbial collaborations. Second, crop breeding is typically conducted in environmental backgrounds receiving high levels of inputs, sometimes even greater than is economically viable for farmers (Boyer 1982). This has led to the selection of cultivars that are high yielding and dependent on these inputs. Breeding programs that select for plant consortia under in reduced input environments where internal ecosystem processes are enhanced will result in biotic assemblages which are adapted to these conditions.

Limited efforts have applied interdisciplinary approaches to crop breeding that combine plant selection for multiple traits (including those related to below-ground functions) with use of reduced input environments (Banziger and Cooper 2001; Alves *et al.* 2003). A successful example of this approach is the soybean breeding program in Brazil where N-fertilizers have been omitted from the soybean breeding programs since the 1960s (reviewed by Alves *et al.* 2003). More efficient *Bradyrhizobium* strains that support higher levels of biological nitrogen fixation (BNF) through the symbiosis were introduced and at first failed to compete against indigenous strains for nodule space (Nishi *et al.* 1996). Over 10 years as these strains adapted to the soil environment, a few of the introduced strains developed improved competitive ability against less-efficient indigenous strains for nodule space. One of these strains is now commonly used in the surrounding region as an inoculant in soybean. This observation suggests that selecting for desired plant traits under appropriate environmental conditions (in this case, improved N-fixing mutualism selected for under conditions of low soil N) has cascading effects on symbiont populations and selects for plant–microbial associations that function well in agricultural systems.

For the most part, selecting for cash crop species that are well suited for ecologically complex production systems will require that yield-related traits (quantity and quality of harvestable crop) continue to be of primary importance while selection for specific rhizosphere functions serves a supporting role. In contrast, rhizosphere function can serve as a primary trait in breeding for cover crops which support specific ecosystem services that are mediated by plant–microbial interaction in the rhizosphere. We believe that the development of plant–microbial consortia that can enhance specific ecosystem processes such as aggregation, nutrient bioavailability and retention, disease suppression and C-storage is a reasonable goal.

USING INCREASED PLANT SPECIES BIODIVERSITY IN SPACE AND TIME

Plant species diversity can be increased either by introducing additional cash crops or non-cash crops (hereafter referred to auxiliary crops, cover crops or intercrops) selected to serve specific ecosystem functions. The benefits of management practices that expand plant presence in space and time have long been recognized (Hambridge 1938). Given the current understanding of belowground plant functions, intentional management of plant diversity based on the capacity of a species to enhance particular ecosystem processes is clearly feasible. Indeed, the potential for a single plant species and its associated microbes to significantly influence ecosystem function is large in agroecosystems since single species effects tend to be more pronounced in ecosystems with limited biodiversity (Chapin *et al.* 2000), particularly when a missing functional group is added (Naeem and Li 1997).

The prospect of increasing rotational diversity and replacing bare fallows with cover crops is generally deemed impractical mainly due to concerns about yield reductions, and the bulk of research evaluating diversified rotations has focused solely on yield assessments (Tonitto *et al.* 2006). There are several recent examples in which assessment expands to include the potential contributions of cover crops to a wide range of belowground agroecosystem services including aggregation (Haynes and Beare 1997), creation of biopores in compacted soils (Williams and Wiel 2004), P bioavailability (Oberson *et al.* 1999), disease suppression (Mazzola 2004), and N use efficiency (Tonitto *et al.* 2006). A recent meta-analysis of the cover crop literature suggests that yield penalties have been over emphasized in the past (Tonitto *et al.* 2006). Yields in diversified rotations employing cover crops varied from no detectable yield reductions to an average yield reduction of 10 percent in studies where corn was grown following a leguminous cover crop without added fertilizer N. Nitrate leaching, which was the only ecosystem service that had been sufficiently studied to include in the meta-analysis, was reduced by 70 percent on average. Further assessments of cropping systems where restored plant diversity replaces conventional inputs to provide ecosystem services are needed. Identifying a wider variety of species as well as modifying current cover crop species through breeding programs to fill particular niches could also greatly increase the potential for cover crop adoption and expand contributions from rhizosphere processes.

INTEGRATED PLANT AND AMENDMENT STRATEGIES

Ecosystem-based approaches to soil fertility management targeting soil nutrient reservoirs with longer mean residence times that can be accessed by plants and their associated microbes has the potential to build soil productivity over time (Drinkwater and Snapp 2007). Integrated management of biogeochemical processes that regulate the cycling of nutrients and carbon, combined with increased reservoirs more readily retained in the soil, will greatly reduce the need for surplus nutrient additions in agriculture. This approach relies on numerous rhizosphere processes, and combines the use of organic amendments and small amounts of inorganic nutrients from sparingly soluble sources such as rock phosphate with inclusion of plant consortia that can access these sources. Greater understanding of the ecology and evolution of the microbial loop in the rhizosphere will be crucial to coupling N-mineralization with plant uptake so that N-losses and plant–microbial competition for N are minimized.

Strategies need to be developed that combine P application with assimilation in biological sinks, through management and integration of species that augment levels of soil organic acids and phosphates. Application of sparingly soluble sources of P to crops (e.g., most legumes) that can assimilate P into biological pools is an efficient strategy that has been underappreciated, and could be used to bypass desorption, precipitation and occlusion of P (Vance *et al.* 2003). Legumes are important vehicles to enhance P availability through diverse mechanisms, including modified roots, secretion of organic acids and enhanced P-solubilizing activity through microorganisms (Oberson *et al.* 1999). Similarly, targeted use of animal manures can facilitate plant and microbial uptake of P and enhance crop access to P (Erich *et al.* 2002). Manipulation of mycorrhizal populations to develop more efficient plant–symbiont combinations is in its infancy, but strategies that can be pursued include use of sparingly soluble rock P, reduced tillage and integration of auxiliary plants that are highly mycorrhizal.

CONCLUSIONS

Throughout this chapter we have emphasized interactions between the managed soil environment and the biota inhabiting plant-associated niches. Improved ecosystem-based strategies will require an understanding of the feedbacks among management, rhizosphere communities and the background soil environment at longer-temporal scales than current investigations tend to encompass. In particular, future research must address feedbacks between abiotic conditions and ecological and evolutionary processes that govern rhizosphere community structure and function. We believe that adoption of the ecosystem as a conceptual model will guide future research in these directions.

We recognize that ultimately the transition to ecologically sound, sustainable food production systems that meet human needs will be complex and will require fundamental changes in cultural values and human societies (Boyden 2004) as well as the application of ecological knowledge to agricultural management. It is our hope that application of current ecological understanding to the design of agricultural systems will provide the scientific know-how to promote the transition to sustainable food systems that supply a wide range of necessary ecosystem services. We believe there is a tremendous untapped potential for subterranean ecological processes to contribute to these sustainable food systems.

REFERENCES

Abawi, G.S., and T.L. Widmer. 2000. Impact of soil health management practices on soilborne pathogens, nematodes and root diseases of vegetable crops. *Applied Soil Ecology* 15:37–47.
Achouak, W., R. Christen, M. Barakat, M.H. Martel, and T. Heulin. 1999. *Burkholderia caribensis* sp. nov., an exopolysaccharide-producing bacterium isolated from vertisol microaggregates in Martinique. *International Journal of Systematic Bacteriology* 49:787–794.

Alami, Y., W. Achouak, C. Marol, and T. Heulin. 2000. Rhizosphere soil aggregation and plant growth promotion of sunflowers by an exopolysaccharide-producing *Rhizobium* sp strain isolated from sunflower roots. *Applied and Environmental Microbiology* 66:3393–3398.

Alves, B.J.R., R.M. Boddey, and S. Urquiaga. 2003. The success of BNF in soybean in Brazil. *Plant and Soil* 252:1–9.

Angers, D.A., and J. Caron. 1998. Plant-induced changes in soil structure: Processes and feedbacks. *Biogeochemistry* 42:55–72.

Baer, S.G., D.J. Kitchen, J.M. Blair, and C.W. Rice. 2002. Changes in ecosystem structure and function along a chronosequence of restored grasslands. *Ecological Applications* 12:1688–1701.

Banziger, M., and M. Cooper. 2001. Breeding for low input conditions and consequences for participatory plant breeding: Examples from tropical maize and wheat. *Euphytica* 122:503–519.

Bertholdsson, N.O. 2004. Variation in allelopathic activity over 100 years of barley selection and breeding. *Weed Research* 44:78–86.

Bossio, D.A., M.S. Girvan, L. Verchot, J. Bullimore, T. Borelli, A. Albrecht, K.M. Scow, A.S. Ball, J.N. Pretty, and A.M. Osborn. 2005. Soil microbial community response to land use change in an agricultural landscape of Western Kenya. *Microbial Ecology* 49:50–62.

Boyden, S. 2004. *Biology of Civilization: Understanding Human Culture as a Force in Nature.* University of New South Wales Press, Sydney, Australia. 189 pp.

Boyer, J.S. 1982. Plant productivity and environment. *Science* 218:443–448.

Briones, A.M., S. Okabe, Y. Umemiya, N.B. Ramsing, W. Reichardt, and H. Okuyama. 2002. Influence of different cultivars on populations of ammonia-oxidizing bacteria in the root environment of rice. *Applied and Environmental Microbiology* 68:3067–3075.

Briones, A.M. Jr., S. Okabe, Y. Umemiya, N.B. Ramsing, W. Reichardt, and H. Okuyama. 2003. Ammonia-oxidizing bacteria on root biofilms and their possible contribution to N use efficiency of different rice cultivars. *Plant and Soil* 250:335–348.

Buckley, D.H., and T.M. Schmidt. 2003. Diversity and dynamics of microbial communities in soils from agro-ecosystems. *Environmental Microbiology* 5:441.

Buenemann, E.K., D.A. Bossio, P.C. Smithson, E. Frossard, and A. Oberson. 2004. Microbial community composition and substrate use in a highly weathered soil as affected by crop rotation and P fertilization. *Soil Biology and Biochemistry* 36:889–901.

Buyanovsky, G.A., and C.L. Kucera. 1987. Comparative analyses of carbon dynamics in native and cultivated ecosystems. *Ecology* 68:2023–2031.

Chapin, F.S., III, E.S. Zavaleta, V.T. Eviner, R.L. Naylor, P.M. Vitousek, H.L. Reynolds, D.U. Hooper, S. Lavorel, O.E. Sala, S.E. Hobbie, M.C. Mack, and S. Diaz. 2000. Consequences of changing biodiversity. *Nature* 405:234–242.

Clarholm, M. 1985. Interactions of bacteria, protozoa and plants leading to mineralization of soil nitrogen. *Soil Biology Biochemistry* 17:181–187.

Cook, R.J. 1993. Making greater use of introduced microorganisms for biological-control of plant-pathogens. *Annual Review of Phytopathology* 31:53–80.

da Mota, F.F., A. Nobrega, I.E. Marriel, E. Paiva, and L. Seldin. 2002. Genetic diversity of *Paenibacillus polymyxa* populations isolated from the rhizosphere of four cultivars of maize (*Zea mays*) planted in Cerrado soil. *Applied Soil Ecology* 20:119–132.

da Silva, K.R.A., J.F. Salles, L. Seldin, and J.D. van Elsas. 2003. Application of a novel *Paenibacillus*-specific PCR-DGGE method and sequence analysis to assess the diversity of *Paenibacillus* spp. in the maize rhizosphere. *Journal of Microbiological Methods* 54:213–231.

Dale, V.H., S. Brown, R.A. Haeuber, N.T. Hobbs, N. Huntly, R.J. Naiman, W.E. Riebsame, M.G. Turner, and T.J. Valone. 2000. Ecological principles and guidelines for managing the use of land. *Ecological Applications* 10:639–670.

Dawkins, R. 1982. *Extended Phenotype: The Gene as the Unit of Selection.* Oxford Press, San Francisco.

Devare, M.H., C.M. Jones, and J.E. Thies. 2004. Effect of Cry3Bb transgenic corn and tefluthrin on the soil microbial community: Biomass, activity, and diversity. *Journal of Environmental Quality* 33:837–843.

Douds, D., L. Galvez, R.R. Janke, and P. Wagoner. 1995. Effects of tillage and farming system upon populations and distribution of versicular-arbuscular mycorrhizal fungi. *Agriculture Ecosystems and Environment* 52:111–118.

Drinkwater, L.E. 2004. Improving fertilizer nitrogen use efficiency through an ecosystem-based approach. In: Mosier, A., Syers, J.K., and Freney, J. (eds) *Agriculture and the Nitrogen Cycle: Assessing the Impacts of Fertilizer Use on Food Production and the Environment*, Vol 65. Island Press, Washington, DC, pp. 93–102.

Drinkwater, L.E., and S.S. Snapp. 2007. Nutrients in Agroecosystems: Rethinking the Management Paradigm. *Advances in Agronomy* 92:163–186.

Dunfield, K.E., and J.J. Germida. 2004. Impact of genetically modified crops on soil- and plant-associated microbial communities. *Journal of Environmental Quality* 33:806–815.

Erich, M.S., C.B. Fitzgerald, and G.A. Porter. 2002. The effect of organic amendments on phosphorus chemistry in a potato cropping system. *Agriculture Ecosystems and Environment* 88: 79–88.

Espinosa-Victoria, D., C.P. Vance, and P.H. Graham. 2000. Host variation in traits associated with crown nodule senescence in soybean. *Crop Science* 40:103–109.

Friedland, W., L. Busch, F. Buttel, and A. Rudy. 1991. *Towards a New Political Economy of Agriculture*. Westview Press, Boulder.

Gahoonia, T.S., N.E. Nielsen, and O.B. Lyshede. 1999. Phosphorus (P) acquisition of cereal cultivars in the field at three levels of P fertilization. *Plant and Soil* 211:269–281.

Galvez, L., D.D. Douds, Jr., L.E. Drinkwater, and P. Wagoner. 2001. Effect of tillage and farming system upon VAM fungus populations and mycorrhizas and nutrient uptake of maize. *Plant and Soil* 228:299–308.

Gardener, B.B.M., and D.M. Weller. 2001. Changes in populations of rhizosphere bacteria associated with take-all disease of wheat. *Applied and Environmental Microbiology* 67:4414–4425.

Germida, J.J., and S.D. Siciliano. 2001. Taxonomic diversity of bacteria associated with the roots of modern, recent and ancient wheat cultivars. *Biology and Fertility of Soils* 33:410–415.

Gouzou, L., G. Burtin, R. Philippy, F. Bartoli, and T. Heulin. 1993. Effect of inoculation with *Bacillus polymyxa* on soil aggregation in the wheat rhizosphere – preliminary examination. *Geoderma* 56:479–491.

Guemouri-Athmani, S., O. Berge, M. Bourrain, P. Mavingui, J.M. Thiery, T. Bhatnagar, and T. Heulin. 2000. Diversity of *Paenibacillus polymyxa* populations in the rhizosphere of wheat (*Triticum durum*) in algerian soils. *European Journal of Soil Biology* 36:149–159.

Hambridge, G. 1938. Soils and men- A summary. In: *Soils and Men, Yearbook of Agriculture*. United States Government Printing Office, Washington, DC, pp. 1–44.

Hamel, C. 1996. Prospects and problems pertaining to the management of arbuscular mycorrhizae in agriculture. *Agriculture Ecosystems and Environment* 60:197–210.

Haynes, R.J., and M.H. Beare. 1997. Influence of six crop species on aggregate stability and some labile organic matter fractions. *Soil Biology and Biochemistry* 29:1647–1653.

Heulin, T., O. Berge, P. Mavingui, L. Gouzou, K.P. Hebbar, and J. Balandreau. 1994. *Bacillus polymyxa* and *Rahnella aquatilis*, the dominant N2-fixing bacteria associated with wheat rhizosphere in French soils. *European Journal of Soil Biology* 30:35–42.

Hoitink, H.A.J., and J. Boehm. 1999. Biocontrol within the context of soil microbial communities: A substrate-dependent phenomenon. In: Webster, R.K. (ed.) *Annual Review of Phytopathology*, Vol 37. Annual Reviews Inc., Palo Alto, pp. 427–446.

Hultberg, M., B. Alsanius, and P. Sundin. 2000. *In vivo* and *in vitro* interactions between *Pseudomonas fluorescens* and *Pythium ultimum* in the suppression of damping-off in tomato seedlings. *Biological Control* 19:1–8.

International Federation of Organic Agriculture Movements. 2005. In. http:/www.ifoam.org

Jackson, L.E. 1995. Root architecture in cultivated and wild lettuce (*Lactuca* spp.). *Plant Cell and Environment* 18:885–894.

Jackson, L.E., and G.W. Koch. 1997. The ecophysiology of crops and their wild relatives. In: Jackson, L.E. (ed.) *Ecology in agriculture*. Academic Press Inc., San Diego, CA, pp. 3–37.

Jasper, D.A., A.D. Robson, and L.K. Abbott. 1979. Phosphorus and the formation of vesicular-arbuscular mycorrhizas. *Soil Biology and Biochemistry* 11:501–505.

Jeffries, P., S. Gianinazzi, S. Perotto, K. Turnau, and J.M. Barea. 2003. The contribution of arbuscular mycorrhizal fungi in sustainable maintenance of plant health and soil fertility. *Biology and Fertility of Soils* 37:1–16.

Johnson, M.T.J., and A.A. Agrawal. 2005. Plant genotype and environment interact to shape a diverse arthropod community on evening primrose (*Oenothera biennis*). *Ecology* 86:874–885.

Kiers, E.T., S.A. West, and R.F. Denison. 2002. Mediating mutualisms: Farm management practices and evolutionary changes in symbiont co-operation. *Journal of Applied Ecology* 39:745–754.

Kumar, B.S.D., I. Berggren, and A.M. Martensson. 2001. Potential for improving pea production by co-inoculation with fluorescent *Pseudomonas* and *Rhizobium*. *Plant and Soil* 229:25–34.

Lewis, W.J., L.J.C. Van, S.C. Phatak, and J.H. Tumlinson, III. 1997. A total system approach to sustainable pest management. *Proceedings of the National Academy of Sciences of the United States of America* 94:12243–12248.

Lynch, J.M. 1990. *The Rhizosphere*. John Wiley & sons, West Sussex.

Mazzola, M. 2004. Assessment and management of soil microbial community structure for disease suppression. *Annual Review of Phytopathology* 42:35–59.

Mazzola, M., D.L. Funnell, and J.M. Raaijmakers. 2004. Wheat cultivar-specific selection of 2,4-diacetylphloroglucinol-producing fluorescent *Pseudomonas* species from resident soil populations. *Microbial Ecology* 48:338–348.

McGonigle, T.P., M.H. Miller, and D. Young. 1999. Mycorrhizae, crop growth, and crop phosphorus nutrition in maize-soybean rotations given various tillage treatments. *Plant and Soil* 210:33–42.

Moawad, H., S.M.S.B. El Din, and R.A. Abdel Aziz. 1998. Improvement of biological nitrogen fixation in Egyptian winter legumes through better management of *Rhizobium*. *Plant and Soil* 204:95–106.

Morris, C.E., and J.M. Monier. 2003. The ecological significance of biofilm formation by plant-associated bacteria. *Annual Review of Phytopathology* 41:429–453.

Naeem, S., and S. Li. 1997. Biodiversity enhances ecosystem reliability. *Nature London* 390:507–509.

National Organic Program. 2005. United States Department of Agriculture. In. http:/www.ams.usda.gov/nop/NOP/standards/ProdHandReg.html

Neff, J.C., A.R. Townsend, G. Gleixner, S.J. Lehman, J. Turnbull, and W.D. Bowman. 2002. Variable effects of nitrogen additions on the stability and turnover of soil carbon. *Nature London* 419:915–917.

Nelson, A.D., L.E. Barber, J. Tjepkema, S.A. Russell, R. Powelson, H.J. Evans, and R.J. Seidler. 1976. Nitrogen-fixation associated with grasses in Oregon. *Canadian Journal of Microbiology* 22:523–530.

Nishi, C.Y.M., L.H. Boddey, M.A.T. Vargas, and M. Hungria. 1996. Morphological, physiological and genetic characterization of two new *Bradyrhizobium* strains recently recommended as Brazilian commercial inoculants for soybean. *Symbiosis* 20:147–162.

Oberson, A., D.K. Friesen, H. Tiessen, C. Morel, and W. Stahel. 1999. Phosphorus status and cycling in native savanna and improved pastures on an acid low-P Colombian Oxisol. *Nutrient Cycling in Agroecosystems* 55:77–88.

Ott, S. 1990. Supermarket shoppers' pesticide concerns and willingness to purchase certified pesticide residue-free fresh produce. *Agribusiness* 6:593–602.

Paterson, E. 2003. Importance of rhizodeposition in the coupling of plant and microbial productivity. *European Journal of Soil Science* 54:741–750.

Pinton, R., Z. Varanini, and P. Nannipieri. (eds) 2001. *The Rhizosphere: Biochemistry and Organic Substances at the Soil-Plant Interface.* Marcel Dekker, New York.

Pryor, F.L. 1985. The invention of the plow. *Comparative Studies in Society and History* 27:727–743.

Reganold, J.P., J.D. Glover, P.K. Andrews, and H.R. Hinman. 2001. Sustainability of three apple production systems. *Nature* 410:926–930.

Reid, J.B., and M.J. Goss. 1981. Effect of living roots of different plant species on the aggregate stability of 2 arable soils. *Journal of Soil Science* 32:521–542.

Rengel, Z. 2002. Breeding for better symbiosis. *Plant and Soil* 245:147–162.

Rillig, M.C., and P.D. Steinberg. 2002. Glomalin production by an arbuscular mycorrhizal fungus: A mechanism of habitat modification? *Soil Biology and Biochemistry* 34:1371–1374.

Rillig, M.C., S.F. Wright, K.A. Nichols, W.F. Schmidt, and M.S. Torn. 2001. Large contribution of arbuscular mycorrhizal fungi to soil carbon pools in tropical forest soils. *Plant and Soil* 233:167–177.

Rillig, M.C., S.F. Wright, and V.T. Eviner. 2002. The role of arbuscular mycorrhizal fungi and glomalin in soil aggregation: Comparing effects of five plant species. *Plant and Soil* 238: 325–333.

Robleto, E.A., J. Borneman, and E.W. Triplett. 1998. Effects of bacterial antibiotic production on rhizosphere microbial communities from a culture-independent perspective. *Applied and Environmental Microbiology* 64:5020–5022.

Salles, J.F., J.A. van Veen, and J.D. van Elsas. 2004. Multivariate analyses of *Burkholderia* species in soil: Effect of crop and land use history. *Applied and Environmental Microbiology* 70:4012–4020.

Schloter, M., M. Lebuhn, T. Heulin, and A. Hartmann. 2000. Ecology and evolution of bacterial microdiversity. *Fems Microbiology Reviews* 24:647–660.

Siciliano, S.D., J.J. Germida, K. Banks, and C.W. Greer. 2003. Changes in microbial community composition and function during a polyaromatic hydrocarbon phytoremediation field trial. *Applied and Environmental Microbiology* 69:483–489.

Smith, R.F., and R. van der Bosch. 1967. Integrated Control. In: Kilgore, W.W., and Doutt, R.L. (eds) *Pest Control: Biological, Physical, and Selected Chemical Methods.* Academic Press, New York, pp. 295–340.

Snapp, S.S., C. Shennan, and A.H.C. Vanbruggen. 1991. Effects of salinity on severity of infection by *Phytophthora parasitica* Dast, ion concentrations and growth of tomato, *Lycopersicon esculentum* Mill. *New Phytologist* 119:275–284.

Snapp, S., W. Kirk, B. Roman-Aviles, and J. Kelly. 2003. Root traits play a role in integrated management of *Fusarium* root rot in snap beans. *Hortscience* 38:187–191.

Spaink, H.P., A. Kondorosi, and P.J. Hooykaas. (eds) 1998. The Rhizobiaceae: Molecular Biology of Model Plant-Associated Bacteria, 13th edn. Kluwer Acedemic, Boston.

Sturz, A.V., and B.R. Christie. 2003. Beneficial microbial allelopathies in the root zone: The management of soil quality and plant disease with rhizobacteria. *Soil & Tillage Research* 72:107–123.

Tilman, D. 1999. Global environmental impacts of agricultural expansion: The need for sustainable and efficient practices. *Proceedings of the National Academy of Sciences of the United States of America* 96:5995–6000.

Tisdall, J.M., and J.M. Oades. 1982. Organic matter and water-stable aggregates in soils. *Journal of Soil Science* 33:141–163.

Tonitto, C., M. David, and L.E. Drinkwater. 2006. Replacing bare fallows with cover crops in fertilizer-intensive cropping systems: A meta-analysis of crop yield and N dynamics. *Agriculture Ecosystems and Environment* 112:58–72.

Topp, G.C., K.C. Wires, D.A. Angers, M.R. Carter, J.L.B. Culley, D.A. Holmstrom, B.D. Kay, G.P. LaFond, D.R. Langille, R.A. McBride, G.T. Patterson, E. Perfect, V. Rasiah, A.V. Rodd, and K.T. Webb. 1995. Changes in soil stucture. In: Acton, D.F., and Gregorich, L.J. (eds) *The Health of our Soils-Toward Sustainable Agriculture in Canada.* Centre for Land and Biological Resources Research, Research Branch, Agriculture and Agri-Food Canada, Ottawa, Ont., pp. 51–60.

Vance, C.P., C. Uhde-Stone, and D.L. Allan. 2003. Phosphorus acquisition and use: Critical adaptations by plants for securing a nonrenewable resource. *New Phytologist* 157:423–447.

van Wees, S.C.M., M. Luijendijk, I. Smoorenburg, L.C. van Loon, and C.M.J. Pieterse. 1999. Rhizobacteria-mediated induced systematic resistance (SIR) in *Arabidopsis. Plant Molecular Biology* 41:537–549.

Vitousek, P.M., J.D. Aber, R.H. Howarth, G.E. Likens, P.A. Matson, D.W. Schindler, W.H. Schlesinger, and D.G. Tilman. 1997a. Human alteration of the global nitrogen cycle: Source and consequences. *Ecological Applications* 7:737–750.

Wander, M.M. 2004. Soil organic matter fractions and their relevance to soil function. In: Magdoff, F., and Weil, R.R. (eds) *Soil Organic Matter in Sustainable Agriculture*, Vol 11. CRC Press LLC, Boca Raton, pp. 67–102.

Whipps, J.M. 2001. Microbial interactions and biocontrol in the rhizosphere. *Journal of Experimental Botany* 52:487–511.

Whitham, T.G., W.P. Young, G.D. Martinsen, C.A. Gehring, J.A. Schweitzer, S.M. Shuster, G.M. Wimp, D.G. Fischer, J.K. Bailey, R.L. Lindroth, S. Woolbright, and C.R. Kuske. 2003. Community and ecosystem genetics: A consequence of the extended phenotype. *Ecology* 84:559–573.

Williams, S.M., and R.R. Weil. 2004. Crop cover root channels may alleviate soil compaction effects on soybean crop. *Soil Science Society of America Journal* 68:1403–1409.

Wright, S.F., M. FrankeSnyder, J.B. Morton, and A. Upadhyaya. 1996. Time-course study and partial characterization of a protein on hyphae of arbuscular mycorrhizal fungi during active colonization of roots. *Plant and Soil* 181:193–203.

Wright, S.F., and R.L. Anderson. 2000. Aggregate stability and glomalin in alternative crop rotations for the central great plains. *Biology and Fertility of Soils* 31:249–253.

CHAPTER 7

The Contribution of Root – Rhizosphere Interactions to Biogeochemical Cycles in a Changing World

Kurt S. Pregitzer, Donald R. Zak, Wendy M. Loya, Noah J. Karberg, John S. King, and Andrew J. Burton

7.1 INTRODUCTION

In this chapter, we will discuss how elevated atmospheric CO_2 and atmospheric N deposition influence the physiology and growth of fine roots, as well as how changes in root system form and function can influence rhizosphere processes, which scale-up to alter biogeochemical cycles. Our intent is to review some of the processes that we feel are most important in understanding ecosystem feedbacks. We do not attempt to review all the literature in this important area, nor do we attempt to discuss all relevant topics. We draw heavily from our own research, presenting important examples of how rhizosphere interactions can influence ecosystem-level feedbacks. Despite the limitations in the scope of this chapter, we hope our treatment of this subject stimulates further research into the mechanisms controlling ecosystem function as the Earth's atmosphere changes.

The concentration of carbon dioxide in Earth's atmosphere is increasing rapidly due to human combustion of fossil fuels and changes in land use. Decades of experimentation using growth chambers, glasshouses, open-top chambers, and now FACE experiments provide evidence that rising atmospheric CO_2 will increase primary productivity in the absence of strong limitation by other resources (Ceulemans and Mousseau 1994; Curtis 1996; Curtis and Wang 1998; DeLucia *et al.* 1999). Atmospheric N deposition has also greatly increased in terrestrial ecosystems, again due to human activities (Galloway 1995; Vitousek *et al.* 1997). Soil N availability influences plant tissue N concentration, which plays a key role in regulating leaf-level photosynthesis (Field and Mooney 1986; Reich *et al.* 1997), and rates of tissue respiration per unit mass (Ryan 1991). Soil N availability can also influence C allocation to root systems and rates of root turnover (Nadelhoffer *et al.* 1985; Pregitzer *et al.* 1995; Burton *et al.* 2000), mineralization of C and N by soil microbes in the rhizosphere (Zak and Pregitzer 1998), and the flux of C back to the atmosphere via soil respiration (Burton *et al.* 2004). Thus, both atmospheric CO_2 and N deposition can impact biogeochemical cycles by directly altering net primary productivity (NPP).

In terrestrial ecosystems, plants are the transducers that provide the energy for microbial metabolism through root exudation, cell sloughing, and root and mycorrhizal turnover. Increasing atmospheric CO_2 and N deposition will modify NPP and plant C allocation, and this will, in turn, initiate a series of biochemical changes in dead leaves and fine roots, a response which moves through the rhizosphere to structure soil food webs and control rates of ecosystem C- and N-cycling (Figure 7.1). In our conceptual model, fine root physiology, tissue biochemistry, and fine root growth and mortality play pivotal functions controlling ecosystem biogeochemistry. In other words, understanding how fine roots and microbial dynamics in the rhizosphere respond to increasing atmospheric CO_2 and N deposition is important to understanding the biogeochemical feedbacks between the C and N cycles, which may ultimately constrain long-term ecosystem responses.

This chapter is organized around Figure 7.1. We will systematically follow the consequences of elevated atmospheric CO_2 and N deposition on root growth, turnover, respiration, and tissue biochemistry. We will then briefly discuss how altered root function will influence microbial community composition and function in the rhizosphere. We demonstrate how altered root physiology and growth can be coupled with an understanding of microbial bioenergetics in the rhizosphere to produce a conceptual framework that will help in predicting ecosystem feedback and long-term patterns of ecosystem C and N cycling. Finally, we will explore how roots and microbes are linked to ecosystem-level feedbacks like soil respiration and dissolved inorganic carbon (DIC) leaching (Figure 7.1). Our basic premise is that altered atmospheric chemistry directly impacts fine root form and function, and that human-induced changes to the Earth's atmosphere will cascade through plant root systems into the rhizosphere, where microbial communities in soil mediate the ecosystem feedbacks that regulate the cycling of C and N.

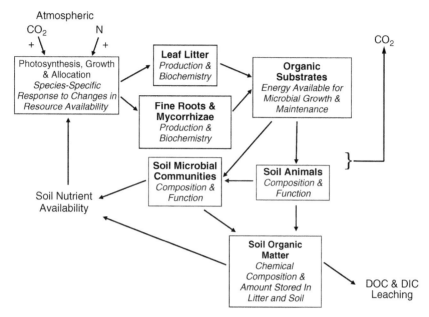

FIGURE 7.1 A model of ecosystem feedback under elevated atmospheric CO_2 and increasing N deposition. Ecosystem feedback is initiated by the response of dominant plant species to changing resource availability (i.e., C acquisition under CO_2 and N deposition). Species-specific responses in terms of primary growth and secondary tissue biochemistry directly modify the quantity and biochemical composition of plant litter, which, in turn, controls the energy soil microorganisms and animals can obtain during litter decomposition. Substrates favoring growth stimulate a microbial demand for N, potentially decreasing the amount available to plants and increasing amounts incorporated into soil organic matter. Thus, we predict that changes in soil N availability, which feed back to control plant growth response to elevated atmospheric CO_2, is regulated by microbial bioenergetics and amounts of N required for microbial biosynthesis. This interaction between elevated atmospheric CO_2 and soil N availability as impacted by atmospheric N deposition is still poorly understood.

7.2 A CONCEPTUAL OVERVIEW OF RHIZOSPHERE PROCESSES IN BIOGEOCHEMICAL CYCLES

Our conceptual model is based on the idea that altered resource availability (elevated atmospheric CO_2, increased atmospheric N deposition) modifies plant C allocation and initiates a series of physiological and biochemical changes in live and dead leaves and fine roots, a response that structures rhizosphere food webs and controls rates of soil C and N cycling (Figure 7.1; see Chapter 5). In Figure 7.1, leaf and fine root litter production and biochemistry play a central role in controlling the flow of energy and N through trophic levels in soil. Litter biochemistry and production are key to understanding processes and the biogeochemical feedbacks between the C and N cycles, which may ultimately constrain the long-term response of ecosystems to rising atmospheric CO_2 or increases in atmospheric N deposition. A better understanding of the physiological links among plant C allocation, litter biochemistry, and rhizosphere food web response is important for several reasons. Central to predicting whether initial enhancement of NPP under elevated CO_2 will decline is understanding if soil N availability will increase, decline, or remain unchanged. Although many investigators have sought to understand how soil N availability will change under elevated CO_2, responses in the literature are highly variable, and currently there is no theoretical or conceptual framework that can explain why soil N availability increases in some studies and declines in others (Zak et al. 2000a). We propose that an understanding of how changing resource availability influences the allocation of photosynthate to growth, storage, and defense in all plant tissues, coupled with an understanding of microbial bioenergetics, holds the promise of providing a conceptual framework which will enable us to predict ecosystem feedback and long-term patterns of ecosystem C and N cycling as CO_2 accumulates in the Earth's atmosphere and ecosystems approach N saturation. We limit our discussion to the role of fine roots in rhizosphere processes and biogeochemical cycles, and attempt to explain how and why elevated atmospheric CO_2 and N deposition alter root physiology, biochemistry, and litter production. These changes, along with those that influence plant shoot systems, cascade through the microbial community to control rhizosphere trophic structure and ecosystem biogeochemistry (Figure 7.1).

FINE ROOT GROWTH

Elevated atmospheric CO_2 increases the growth of small-diameter roots (≤ 1 mm) across a range of species and experimental conditions (Rogers et al. 1994; Pregitzer et al. 1995; Tingey et al. 2000). This response is directly related to the stimulation of photosynthesis under elevated CO_2, and an increase in fine root growth is perhaps the most consistent plant growth response when plants and ecosystems are exposed to elevated CO_2. For example, Norby et al. (2004) found that fine root production more than doubled in elevated CO_2 plots and this response was sustained over 6 years, while other components of tree growth showed little response to elevated CO_2.

The poor understanding of the relationship between fine root biomass and soil N availability is partly due to the inconsistent relationship in literature between soil N availability and root turnover. Atmospheric N deposition also has the potential to alter fine root biomass, but the direction of this response is not as clear as the response to elevated atmospheric CO_2. Reductions in fine root biomass have sometimes occurred as soil N becomes more available (Vogt *et al.* 1990; Haynes and Gower 1995), but this response is not always observed (Burton *et al.* 2004). Reductions in biomass could result from either a decrease in new root growth or an increase in root turnover (shorter mean root lifespan). The lack of a consistent growth response across species and experimental N additions suggests that the life-history attributes of the individual species and/or the symbiotic association of roots with mycorrhizae may be important in understanding fine root growth in response to an increase in soil N availability.

FINE ROOT TURNOVER

Fine root turnover refers to the flux of carbon and nutrients from plants into soil per unit area per unit time, and it is a major component of forest ecosystem carbon and nutrient cycling. Fine root turnover is caused by the production, mortality, and decay of small-diameter roots. Fine root production has been estimated to account for up to 33 percent of global annual NPP (Gill and Jackson 2000), and the lifespan of fine roots is relatively short, so turnover of carbon and nutrients from root mortality is rapid. However, the contribution of fine root turnover to total ecosystem carbon and nutrient budgets remains uncertain because it has always been difficult to directly quantify fine root turnover (Trumbore and Gaudinski 2003).

Soil coring was among the first methods used to estimate the amount of carbon and nutrients cycled via fine root production and decomposition (Nadelhoffer and Raich 1992; Fahey and Hughes 1994). These methods require assumptions about root growth and mortality, which can be difficult to ascertain. The use of minirhizotrons in recent years has improved our knowledge of fine root dynamics, because they allow for the direct observation of fine root production and mortality (Hendrick and Pregitzer 1992; Burton *et al.* 2000; Tierney and Fahey 2001; Ruess *et al.* 2003). However, short-term minirhizotron studies do not adequately track the longevity of longer-lived fine roots, and the tubes themselves can influence root lifespan (Withington *et al.* 2003). Although minirhizotrons have greatly improved our understanding of how fine roots respond to changing soil environments, we still do not understand the mechanisms controlling fine root turnover (Pregitzer 2002).

Many root ecologists have used the simplifying assumption that a fine root is a fine root regardless of the diameter or position of the individual root on the branching root system. Pregitzer *et al.* (2002) showed that specific root length and nitrogen content depend fundamentally on the position of a root on the branching root system. Distal roots have the highest specific root length and nitrogen contents, and, by inference, are more metabolically active. Most fine roots are much smaller than commonly assumed, and species differ in the way in which fine roots are constructed (Pregitzer *et al.* 2002). Another commonly used assumption is that all fine roots within an arbitrary size class or among species have similar rates of turnover, which is not the case (Wells and Eissenstat 2001; King *et al.* 2002; Matamala *et al.* 2003). Roots of the same diameter may have a different branching structure, different function, and different rates of turnover. Guo *et al.* (2004) recently concluded small, fragile, and more easily overlooked first- and second-order roots may be disproportionately important in ecosystem scale C and N fluxes due to their large proportions of fine root biomass, high N concentrations, short lifespans, and potentially high decomposition rates. Therefore, one of our most significant knowledge gaps in understanding how changes in atmospheric chemistry influence biogeochemical cycling appears to be a relatively simple, yet intractable, problem: how to define the pool of carbon and nutrients in fine roots. This simple problem needs to be resolved. In spite of the vague and uncertain nature of how we define root turnover, both elevated atmospheric CO_2 and increasing N deposition have important effects on root turnover.

ELEVATED ATMOSPHERIC CO_2 AND ROOT TURNOVER

Elevated atmospheric CO_2 usually increases absolute rates of fine root turnover $(g\ m^{-2}y^{-1})$, which results in an increased flux of organic substrates and associated nutrients from the root system for microbial growth and maintenance (Figure 7.1). However, most studies suggest that elevated CO_2 does not alter the average lifespan of individual fine roots (Pregitzer *et al.* 1995; Tingey *et al.* 2000). Normally, the increased flux ("turnover") from the roots to the soil is the result of greater fine root production (growth) rather than a change in the lifespan of individual roots. In other words, increased turnover, that is increased rates of C flux from the root system to the soil, is driven primarily by an increase in the rate of root growth under elevated CO_2. One of the more interesting unanswered questions at this time is whether or not increased rates of root growth and absolute turnover normally reported from short-term (1–5 yr) experiments will be sustained as ecosystems continue to be exposed to elevated CO_2 over longer periods of time. It is entirely possible that this trend is a transient response of exposure to elevated atmospheric CO_2, and that root growth will decline if nutrients eventually limit plant growth in response to elevated CO_2 (Oren *et al.* 2001). However, some recent evidence suggest that fine roots respond to elevated CO_2 for many years and grow deeper in the soil, even when stand density is high and shoot growth is unresponsive (Norby *et al.* 2004).

N Deposition and Root Turnover

The effects of increased soil N availability, induced by atmospheric N deposition, on fine root lifespan are not at all clear, in spite of considerable research and years of debate in the literature. In some cases, increased soil N availability clearly decreases mean root lifespan (Pregitzer et al. 1995). In other cases, average root lifespan increases as soil N becomes more available (Burton et al. 2000), or remains unchanged (Burton et al. 2004). The absolute flux of C and N from the root system into the pool of organic substrates available to soil microbes depends, among other things, on both the rate of new root production and the average lifespan of individual fine roots. For example, if rates of production stay the same, but average root lifespan declines as soil N becomes more available, the pool of fine root biomass declines, as does the flux of C and N into the soil. Such a response could result in a decrease in total soil respiration (see below). Until we better understand how soil N availability influences both new root production and the life expectancy of individual roots, it will be impossible to make intelligent generalizations about the turnover of C and N from plant root systems in response to atmospheric N deposition (Trumbore and Gaudinski 2003).

Root Respiration

Root respiration is a major component of total soil CO_2 efflux, usually accounting for at least 50 percent of soil respiration (Hanson et al. 2000). Plant roots utilize photosynthate for maintenance of existing tissue, ion uptake, and new root growth (Eissenstat 1992; Eissenstat and Yanai 1997), and of course fine roots are the carbon depot for mycorrhizae (Leake et al. 2001). Specific rates of root respiration (rate per unit mass), like all plant tissues, depend on tissue N concentration (Pregitzer et al. 1998). As the concentration of tissue N increases, so do rates of respiration. The position of an individual root on the branching root system also influences specific rates of root respiration. Distal roots, which have smaller diameters and higher tissue N concentrations, exhibit much higher rates of respiration per gram of tissue (Pregitzer et al. 1998, 2002; Pregitzer 2002). The fact that the branching root system does not exhibit uniform rates of respiration per unit mass complicates methods of sampling and quantifying root respiration in response to changes in atmospheric CO_2 and N deposition. Nevertheless, several trends are noteworthy.

As explained above, elevated atmospheric CO_2 increases new root growth, and therefore both root respiration (growth and maintenance respiration) and total soil respiration typically increases when ecosystems are exposed to elevated atmospheric CO_2 (King et al. 2001a; see Chapter 2). Again, this response is both logical (driven by changes in root growth) and consistent in the literature.

Experimental additions of N to terrestrial ecosystems normally result in decreased soil respiration (Söderström *et al.* 1983; Haynes and Gower 1995; Homann *et al.* 2001), although positive effects and a lack of response have also been documented (Nohrstedt and Börjesson 1998; Kane *et al.* 2003). Changes in soil CO_2 efflux could result either from reduced respiration of roots and their associated mycorrhizae or from lower microbial respiration. At the ecosystem level, lower root respiration would occur if either specific respiration rates or root biomass declined in response to N additions. Decreases in specific respiration rates are unlikely, because enhanced root N concentrations are associated with greater specific root respiration rates (Ryan *et al.* 1996; Burton *et al.* 1996, 2002). Guo *et al.* (2004) report that N fertilization increased fine root N concentration and content in all five orders of the distal fine root system, resulting in higher specific rates of root respiration. As mentioned above, reductions in root biomass have occurred in response to N fertilization of mature forests (Vogt *et al.* 1990; Haynes and Gower 1995), as the relative amount of C-allocated belowground decreased in response to improved N availability. However, Burton *et al.* (2004) found that after several years of experimentally simulating atmospheric N deposition a significant decrease in soil respiration was not due to reduced allocation of C to roots, as root respiration rates, root biomass, and root turnover were unchanged. It seems clear that field experiments designed to simulate N deposition often result in a decline in soil respiration (Bowden *et al.* 2004; Burton *et al.* 2004), but exactly how or even if root respiration is involved in this response remains unclear. We still do not understand which system-level properties or processes explain the decline in soil respiration often observed in N deposition experiments.

ROOT TISSUE CHEMISTRY

The fate of organic matter in the rhizosphere is influenced by both the quantity produced and its biochemical composition (see Chapter 1). Fine root (and leaf litter) biochemistry is particularly important, because it may directly affect the rate of microbial mediated decomposition of soil organic matter, and tissue biochemistry influences reactions of soil organic matter with the soil mineral particle matrix (Six *et al.* 2002). Currently, very little information exists on changes in fine root biochemistry under elevated CO_2 (Zak *et al.* 2000a). Where data do exist, inconsistencies among experiments make it difficult to place information in a context that is useful for testing mechanistic predictions of organic matter transformations in the soil. Even less is known about the production and biochemistry of fine roots produced under conditions of increasing N deposition. In a manner parallel to work on leaf litter, we now need to determine how chemical changes in root litter will influence decomposer communities. Chemically altered organic inputs could lead to changes in genetic induction of extracellular microbial enzymes or microbial community composition that fundamentally alter rhizosphere C cycling (Larson *et al.* 2002; Phillips *et al.* 2002). An immediate research challenge is to elucidate the roles of quantitative changes in production and qualitative changes in the chemistry of organic inputs from roots to soil and concomitant changes in microbial community composition and metabolism.

Biochemical changes in root tissue, as well as other components of plant litter, will likely be altered in a manner that is consistent with patterns in which plants allocate photosynthate to growth, maintenance, and defense. These, in turn, are tightly linked to life-history traits of the particular plant species involved (Herms and Mattson 1992). Changes in resource availability such as light, nitrogen (N), or atmospheric $[CO_2]$ can affect "growth-dominated" plant species differently than "differentiation-dominated" species. Growth-dominated species tend to allocate "extra" photosynthate (relative to N) to growth, whereas differentiation-dominated species would be expected to allocate "extra" photosynthate to the production of C-based secondary defense compounds (Loomis 1932; Herms and Mattson 1992; Koricheva *et al.* 1998). Although the validity of such "carbon:nutrient balance" or source-sink models has been questioned (Hamilton *et al.* 2001), they have proven useful in developing hypotheses of plant response to changes in resource availability that are consistent with a wide range of observations, and these ideas have guided our thinking about how the changing atmosphere will affect plant growth and C allocation to root growth, storage, and chemical defense.

The carbon:nutrient balance hypothesis (CNBH; Bryant *et al.* 1983) postulates that concentrations of C-based defense compounds increase under conditions favoring carbohydrate accumulation in excess of growth (e.g., elevated CO_2 and high light). The growth-differentiation balance hypothesis (GDBH) states that growth is generally limited by water and nutrients, whereas chemical and morphological differentiation in maturing plant cells depends on available carbohydrates (Lorio 1986; Herms and Mattson 1992). Therefore, production of carbon-based compounds, which defend against herbivory and retard microbial degradation, is enhanced when factors other than photosynthate supply are suboptimal for growth. Both hypotheses provide similar predictions regarding plant C allocation, and they are generally consistent with observed responses to changing C and N availability, at least in green leaves and leaf litter (Koricheva *et al.* 1998; King *et al.* 2001b). Theoretically, elevated CO_2 should increase photosynthate supply in excess of what is required for growth, leading to greater concentrations of root lignin, soluble phenolics, and condensed tannins. Atmospheric N deposition should stimulate shoot sink strength and primary shoot growth, and depress root production and the formation of lignin, soluble phenolics, and condensed tannins in fine roots. In a factorial CO_2 by N experiment, Pregitzer *et al.* (1995) did find that high soil N availability reduced fine root growth and, in a similar experiment, King *et al.* (2005) found increasing soil N availability reduced concentrations of soluble phenolics in fine roots, as expected from C allocation theory. In the same experiment, King *et al.* found that soluble phenolics increased under conditions of elevated atmospheric CO_2. Root lignin concentrations were unresponsive to both elevated atmospheric CO_2 and increasing soil N availability (King *et al.* 2005). Because the inherent life history of plants is so important in regulating primary growth and secondary tissue chemistry, we now need to deliberately study how elevated atmospheric CO_2 and N deposition alter the production of these secondary compounds: compounds which are so important in regulating the composition and function of microbial communities (see Chapter 3).

Non-structural carbohydrates (i.e., simple sugars and starch) and N concentrations in fine roots are also important factors contributing to changes in microbial biosynthesis, N immobilization under elevated atmospheric CO_2, and altered atmospheric N deposition (Zak *et al.* 2000a; see Chapter 2). Non-structural carbohydrates are energy-rich substrates for microbial growth and greater inputs from fine roots should fuel a biosynthetic need for N. Moreover, there is a consistent, negative relationship between the concentration of non-structural carbohydrates and N in most plant tissues. Mooney *et al.* (1995) suggested that this relationship results from the fact that amino acid and starch synthesis compete for a common pool of photosynthate, and, therefore, are mutually exclusive biosynthetic processes. As a consequence, fine root litter with higher concentrations of non-structural carbohydrates should stimulate microbial biosynthesis, but these tissues will likely contain less N to build amino acids, proteins, nucleic acids, and other N-containing compounds in microbial cells. Therefore, increases in non-structural carbohydrate concentrations in plant litter have the potential to greatly stimulate a microbial biosynthetic need for N, leading to higher rates of microbial immobilization in the rhizosphere. The extent to which microbial immobilization would be enhanced should depend on the degree to which non-structural carbohydrate and N concentrations are affected by elevated CO_2 and N deposition. Because both lignin and condensed tannins are energy-poor substrates for microbial growth and do not greatly stimulate microbial biosynthesis, increased concentrations should not foster greater rates of immobilization; however, the degradation of these compounds may form reactive byproducts which lead to the incorporation of N into organic matter via condensation reactions (Stevenson 1994). As such, it appears to be very important to quantify the wide range of compounds in plant litter that affect microbial growth, if we are to predict changes in soil N cycling under elevated CO_2 and increased atmospheric N deposition (see Chapter 3).

We argue that establishing the link between the biochemistry of plant litter inputs to soil and the metabolic response of soil organisms is crucial to a mechanistic understanding of the controls on decomposition and changes in N availability under elevated CO_2 and altered N deposition. Development of a conceptual framework that will allow us to predict ecosystem feedback is important, because changes in soil N availability control the degree to which atmospheric CO_2 and N deposition will stimulate NPP in terrestrial ecosystems (Zak *et al.* 2000b; Oren *et al.* 2001).

7.3   EXAMPLES OF ECOSYSTEM FEEDBACK

Changes in soil nutrient availability in response to an altered atmosphere (elevated atmospheric CO_2, increasing atmospheric N deposition) are strongly influenced by the life history traits of the dominant plant taxa, and the manner in which plants increase or decrease amounts of C and N allocated to plant parts, storage compounds, and defensive chemicals. We propose that altered plant C allocation is the first biological step of a *biochemical signal* that will propagate through ecosystems, affecting soil microbial metabolism and soil nutrient availability in a predictable manner. Depending on how leaf and root litter are altered, changes in production and/or biochemistry will provide either more or less energy for soil microbial metabolism and growth, thereby directly affecting microbial composition and biogeochemical cycling. As outlined in Figure 7.1, understanding how atmospheric chemistry alters the production and biochemistry of plant tissue should help us better understand the feedbacks between rhizosphere processes and a changing atmosphere by elucidating some of the factors that control the decomposition of fine root litter, and therefore the soil's capacity to sequester atmospheric C, and cycle the nutrients essential for terrestrial primary productivity. Below we present three examples from our own research programs to illustrate how important the changes in primary productivity and litter biochemistry can be in regulating ecosystem-level feedbacks.

ATMOSPHERIC N DEPOSITION, LIGNIN DEGRADATION, AND SOIL C STORAGE

In several experiments, we have been investigating the biochemical mechanisms by which atmospheric nitrogen (N) deposition could alter soil C storage. Our rationale for this work was based on the observation that elevated levels of inorganic N in soil solution repress the physiological capacity of white rot basidiomycetes to degrade lignin, a common constituent of the plant cell wall that fosters soil organic matter formation. High levels of inorganic N in soil suppress phenol oxidase and peroxidase, enzymes synthesized by white rot fungi to degrade lignin (Fog 1988). We hypothesized that atmospheric N deposition will foster ecosystem-specific increases and decreases in soil C storage, based on the biochemical constituents of plant litter and the physiological suppression of white rot fungi to high levels of soil N. Such a response should foster an accumulation of soil organic matter in ecosystems dominated by highly lignified litter, whereas greater soil N availability could accelerate the decomposition in ecosystems with low-lignin plant litter in which cellulose degradation is initially N limited (Carreiro et al. 2000). We tested our hypothesis using a field experiment conducted in three types of northern temperate forest that occur across the Upper Lake States region (Zak and Pregitzer 1990). These ecosystems range from 100 percent oak in the overstory (black oak–white oak ecosystem) to 0 percent overstory oak (sugar maple–basswood); the sugar maple–red oak ecosystem has intermediate oak abundance. Thus, these ecosystems span a range of leaf and root litter biochemistry from highly lignified litter (black oak–white oak) to litter with relatively low lignin content (sugar maple–basswood).

In three stands in each ecosystem type, plots were randomly assigned to three levels of atmospheric N deposition (0, 30, and $80 \, kg \, N \, ha^{-1} y^{-1}$). Nitrate is the dominant form of atmospheric N entering these ecosystems (MacDonald et al. 1991), and we imposed our N deposition treatments using six equal increments of $NaNO_3$ applied over the growing season. Our treatments were designed to simulate chronic rates of NO_3^- deposition in the northeastern United States ($30 \, kg \, N \, ha^{-1} y^{-1}$) and Europe ($80 \, kg \, N \, ha^{-1} y^{-1}$; Bredemeier et al. 1998). Shortly following the establishment of this experiment, we have been able to demonstrate that atmospheric N deposition can increase or decrease the soil C storage by modifying the lignolytic capacity of microbial communities in soil (Waldrop et al. 2004).

We hypothesized that atmospheric N deposition would alter soil C storage in an ecosystem-specific manner, the direction of which should depend on the biochemical constituents of the dominant litter and the physiological suppression of lignin degradation by white rot fungi. Our results provide evidence that atmospheric deposition can rapidly and dramatically alter soil C storage in an ecosystem-specific manner (Figure 7.2). We have observed that the addition of $80 \, kg \, NO_3^- \text{-} N y^{-1}$ has produced both significant increases and significant declines in soil C over a relatively short duration. After only 18 months, we observed a 22 percent increase in soil C in the black oak–white oak ecosystem, whereas soil C declined by 24 percent in the sugar maple–basswood ecosystem. Although this rapid response is somewhat surprising, it is consistent with our ideas regarding the suppression of lignin degradation

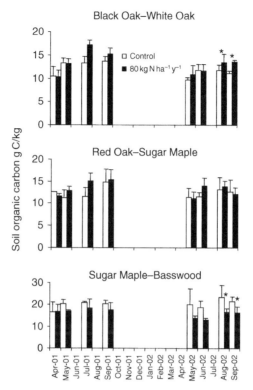

FIGURE 7.2 Soil C in three upland forests receiving experimental atmospheric NO_3^- deposition treatments. The significant increase in soil C in the black oak–white oak and the significant decline in soil C in the sugar maple–basswood ecosystems are consistent with our initial hypothesis. Values are the mean of three stands in each ecosystem type, and the asterisks indicate significant differences between treatment means. After Waldrop *et al.* (2004).

by white rot fungi and its interaction with litter biochemistry. Thus, it appears that atmospheric N deposition could be a global change process that fosters both increases and decreases in soil C storage, depending on the biochemistry of leaf and root litter.

We also observed ecosystem-specific changes in soil enzyme activity, which appear to give rise to the aforementioned changes in soil C storage (compare Figure 7.2 with Figure 7.3). Phenol oxidase and peroxidase activities were significantly enhanced in the sugar maple–basswood ecosystem, we observed no treatment effect in the sugar maple-red oak ecosystem, and the suppression of these enzymes was significant in the black oak–white oak ecosystem (Figure 7.3). This variable response was not consistent with our initial hypothesis, and it presents an interesting opportunity to link microbial

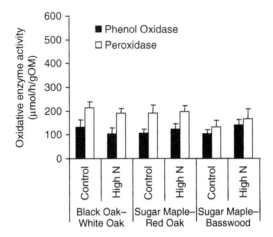

FIGURE 7.3 After 1 year of experimental N addition, we observed an ecosystem-specific response of phenol oxidase and peroxidase. We observed no significant effect of the low N treatment on enzyme activity after 1 year. After Waldrop *et al.* (2004).

community composition and function to ecosystem-level responses to climate change (see Chapter 1). Although high levels of inorganic N can suppress the oxidative capacity of some white rot basidiomycetes (e.g., *Phanerochate chrysosporum*), some evidence in the literature indicates that N deposition can cause an up regulation of lignin oxidation in other soil fungi, even in other genera of white rot basidiomycetes (e.g., *Bjerkandera* and *Trametes*; Collins and Dobson 1997; Hammel 1997). Such a response is evident in the sugar maple–basswood ecosystem (Figure 7.3), in which N deposition *increased* phenol oxidase and peroxidase activity and *decreased* soil C (Figure 7.2). It appears that the response of soil C storage to NO_3^- deposition is controlled by an interaction between microbial community composition/function and litter biochemistry.

The response of soil C storage and phenol oxidase activity is summarized in Figure 7.4; soil C accumulates when phenol oxidase is suppressed and soil C declines when this enzyme is enhanced by N deposition. Fog (1988) argued that the initial states of cellulose degradation were limited by N availability and demonstrated that high levels of inorganic N in soil can stimulate decomposition. Sinsabaugh *et al.* (2002) also have observed that NO_3^- amendment can accelerate decomposition and stimulate cellulase activity in leaf litter. Such a mechanism might also contribute to the overall patterns we have observed, but it is clear that the regulation of phenol oxidase by N availability plays a key role in the loss and accumulation of soil organic matter, and it likely does so in many other terrestrial ecosystems.

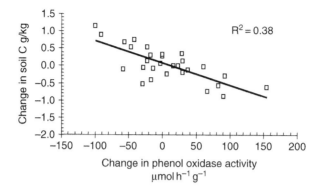

FIGURE 7.4   The relationship between a change in soil C and phenol oxidase activity in control versus $80\,kg\,NO_3{}^{-}\text{-}N\,ha^{-1}\,y^{-1}$ treatments (high N). Reduction in phenol oxidase led to increases in soil C, whereas increases in phenol oxidase reduced soil C. After Waldrop *et al.* 2004.

ALTERED SOIL CARBON UNDER ELEVATED CO₂ AND O₃

The response of ecosystems to elevated atmospheric CO_2 and O_3 provides another interesting example of how altered primary productivity cascades through the root system to influence C storage in the soil. As CO_2 increases in the Earth's atmosphere, so does tropospheric ozone O_3 (Loya *et al.* 2003). Ground-level ozone is a phytotoxin, and large areas of the Earth are exposed to concentrations of tropospheric ozone (O_3) that exceed levels known to be toxic to plants (Chameides *et al.* 1994). In addition to reducing plant growth, exposure to elevated O_3 can also alter plant tissue chemistry (Findlay *et al.* 1996) and reduce allocation of carbon to roots and root exudates (Coleman *et al.* 1995; King *et al.* 1995). In other words, the effects of tropospheric O_3 on root growth are more or less the opposite of those of elevated atmospheric CO_2. The long-term FACE experiment in Rhinelander, Wisconsin, examines how plant–plant and plant–microbe interactions may alter ecosystem responses to elevated O_3 and CO_2 through four treatments: control, elevated CO_2, elevated O_3, and elevated $O_3 + CO_2$. In plots where O_3 and CO_2 are elevated, concentrations were maintained at ~150% of ambient levels (Dickson *et al.* 2000). To examine the effects of atmospheric trace gases on both ecological interactions and on whole ecosystem carbon cycling, each plot is split to include a pure aspen forest, a mixed aspen–birch forest, and a mixed aspen–maple forest. Aspen and birch were chosen because they are among the most widely distributed trees in north temperate forests.

Loya *et al.* (2003) compared soil carbon formation in aspen and aspen–birch subplots under elevated CO_2 and elevated $O_3 + CO_2$ (3 plots each) to understand how exposure to O_3 under elevated CO_2 alters soil carbon formation. They used CO_2 derived from fossil fuel with its highly depleted ^{13}C signature to fumigate plant canopies in the elevated CO_2 and elevated $O_3 + CO_2$ plots. Leaf and root carbon inputs in the elevated CO_2 and elevated $O_3 + CO_2$ rings had a $\delta^{13}C$ signature of $-41.6 \pm 0.4‰$ (mean $\pm$ SE) in contrast to leaf and root inputs of $-27.6 \pm 0.3‰$ in the control rings. The $\delta^{13}C$ signature of soil carbon was $-26.7 \pm 0.2‰$ prior to fumigation and in control plots. Because plant carbon inputs to the soils fumigated with elevated CO_2 and elevated $O_3 + CO_2$ were depleted in ^{13}C relative to soil carbon that existed before the experiment began, incorporation of new root detritus into soil organic matter through time should decrease the $\delta^{13}C$ of soil carbon. Using standard mixing models, they were able to determine the fraction of total soil carbon and acid-insoluble soil carbon derived from atmospheric CO_2 fixed by the trees over the 4 years of experimental fumigation.

Elevated O_3 profoundly altered the ^{13}C composition of soil carbon after 4 years of fumigation. The less depleted $\delta^{13}C$ signature of total soil carbon from the elevated $O_3 + CO_2$ treatment compared to the elevated CO_2 treatment demonstrated that less carbon entered the soil when aspen and mixed aspen–birch forests were exposed to both elevated O_3 and CO_2. Carbon inputs to soil over 4 years accounted for $10.7 \pm 0.6‰$ of the total soil carbon under elevated CO_2, but only $5.7 \pm 0.9\%$ under elevated $O_3 + CO_2$. Thus, elevated O_3 reduced total soil carbon formation by approximately $300\,g\,Cm^{-2}$ compared to the amount formed under elevated CO_2 alone (Figure 7.5).

Soils store organic carbon with a wide range of turnover times. Loya *et al.* (2003) used acid-hydrolysis to isolate the most decay-resistant carbon in the soil (Leavitt *et al.* 1996). The acid-insoluble soil carbon fraction contains the longest-lived compounds, and is comprised primarily of aromatic humic acids

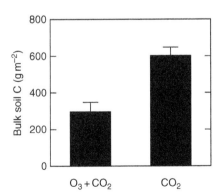

FIGURE 7.5 Total carbon incorporated into soils during 4 years of exposure to elevated $O_3 + CO_2$ and elevated CO_2. Values are means ± 1 SE bars; $P < 0.01$. After Loya *et al.* (2003).

and residues of phenols, lignin, and lignin-associated cellulose from newer plant residues (Paul *et al.* 1997). However, after careful removal of plant residues prior to hydrolysis, new carbon inputs measured in this fraction are restricted to the most decomposition resistant compounds. Compared to elevated CO_2 alone, simultaneous O_3 and CO_2 fumigation greatly reduced the quantity of carbon entering the acid-insoluble fraction in both aspen and aspen–birch soils, as indicated by the less-depleted $\delta^{13}C$ signature (Loya *et al.* 2003). Carbon incorporated into the soil since the initiation of the experiment accounted for approximately 9.1 ± 0.8 percent of the acid-insoluble carbon fraction in the elevated CO_2 plots, but only 4.2 ± 0.6 percent in the elevated $O_3 + CO_2$ plots. Thus, the elevated O_3 treatment reduced formation of acid-insoluble soil carbon by approximately $100 \, g \, Cm^{-2}$ (or 48%) compared with elevated CO_2 alone (Figure 7.6). The atmospheric trace gases CO_2 and O_3 are known to influence photosynthesis and ecosystem productivity in opposite ways. The study by Loya *et al.* (2003) demonstrates how altered atmospheric chemistry has the potential to cascade through plant root systems, altering soil carbon cycling. In response to reduced detrital carbon supply (King *et al.* 2001a ; Percy *et al.* 2002) and increased microbial utilization, forest ecosystems exposed to both elevated O_3 and CO_2 accrued 51 percent less total and 48 percent less acid-insoluble soil carbon, compared to ecosystems exposed only to elevated CO_2. It is still unclear if elevated atmospheric CO_2 will override the negative effects to O_3 on plant growth to increase C storage in soil. More factorial experiments utilizing stable isotope techniques may better help us understand the feedbacks between root growth, litter inputs, and soil C cycling and storage.

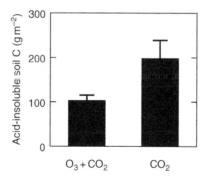

FIGURE 7.6 Carbon incorporated into the stable acid-insoluble fraction of soils during 4 years of exposure to elevated $O_3 + CO_2$ and elevated CO_2. Values are means ± 1 SE bars; $P < 0.01$. After Loya *et al.* (2003).

SOIL RESPIRATION AND THE EXPORT OF DISSOLVED INORGANIC CARBONATE AT ELEVATED CO_2

Our third example of ecosystem feedback links the biotic cycling of C in ecosystems to the export of inorganic C in water leaching from forests. As discussed earlier, atmospheric $[CO_2]$ and leaf tissue N can both directly alter leaf-level photosynthesis, affecting the allocation of photosynthate to root systems and soil respiration. An increase in root growth and soil respiration are two of the most consistent responses in experiments that elevate the concentration of atmospheric CO_2. As CO_2 accumulates in the soil due to the respiration of plant roots and associated soil microorganisms, concentrations of up to 100 times atmospheric levels can result (Berthelin 1988). Soil CO_2 reacts with soil water to form dissolved inorganic carbonate (DIC) through the following reaction:

$$\text{Soil } CO_2 + H_2O = H_2CO_3^* \Leftrightarrow H^+ + HCO_3^- \Leftrightarrow 2H^+ + CO_3^{2-} \qquad (7.1)$$

Thus, changes in belowground carbon allocation driven by changes in leaf-level photosynthesis can significantly control DIC production in the soil.

In experiments that fumigate trees with highly depleted $^{13}CO_2$ (e.g., Loya et al. 2003), a two-member mixing model can be used to better understand the source of increased DIC concentrations in soils. The measured $\delta^{13}C$ of soil pCO_2 and the measured $\delta^{13}C$ of DIC can be used to measure the proportion of DIC-C that is attributable to the fumigation gas. Also, by understanding isotopic fractionation, the seasonal $\delta^{13}C$ of DIC can be used to provide evidence for altered belowground chemistry due to changes in soil CO_2 concentrations. Presumably, changes in the $\delta^{13}C$ ratio of DIC correlate with changes in inorganic chemistry: as soil alkalinity rises and increases the abundance of HCO_3^-, the $\delta^{13}C$ signature of DIC should become increasingly enriched relative to that of soil pCO_2.

Field studies have shown that elevated atmospheric CO_2 increases the concentration of soil CO_2 (Andrews and Schlesinger 2001; King et al. 2001a). Henry's Law (i.e., the amount of gas that dissolves in a liquid is proportional to the partial pressure of the gas over the liquid) dictates that an increase in CO_2 concentration alone will increase the equilibrium concentration of $H_2CO_3^*$ by a factor of $10^{-1.5}$ (Pankow 1991). Thus, not only does elevated soil pCO_2 potentially increase the rate of mineral weathering in soil, it can be expected that elevated atmospheric CO_2 increases the amount of mineral weathering by increasing the formation of $H_2CO_3^*$, that is carbonic acid (Berner 1997; Bormann et al. 1998; Berg and Banwart 2000). Carbonic acid then readily weathers carbonates to form free calcium and bicarbonate. Chapter 8 provides a more detailed description of the processes controlling mineral weathering under increased concentrations of soil CO_2. The combined processes of weathering, riverine export, and deposition of marine carbonates is a transfer and sequestration of atmospheric CO_2, representing a long-term negative feedback to global warming (Drever 1994; Lackner 2002). In other words, there is potentially a direct feedback between increased root growth and root and microbial respiration and DIC export under elevated atmospheric CO_2. However, until recently, the extent to which this feedback mechanism has been fully explored under experimental conditions of elevated atmospheric CO_2 has been limited.

Studying young, fast growing trees under elevated atmospheric CO_2 in a FACE experiment, Karberg et al. (2004) measured a 14 percent increase in soil pCO_2 concentrations, averaged across soil depths and species. Similar increases under elevated CO_2 have previously been reported (Andrews and Schlesinger 2001; King et al. 2001a). With increasing concentrations of soil CO_2, Equation 7.1 is driven to the right, increasing the production of DIC. The increase in DIC was calculated to be 22 percent under elevated CO_2 (Karberg et al. 2004). The concentration of $H_2CO_3^*$ increased by 22 percent due to increased concentrations of soil CO_2, as governed by the Henry's Law Constant, K_H. However, while the concentration of $H_2CO_3^*$ increased under elevated CO_2, it made up a proportionately smaller percentage of DIC-C. The percentage of DIC-C existing as HCO_3^- increased from 11 percent under ambient CO_2 to 17 percent under elevated CO_2. This proportional increase in the bicarbonate ion under elevated CO_2 is accompanied by increases in the concentrations of carbonate (CO_3^{2-}), hydroxide (OH^-), and total alkalinity, as well as a decrease in the concentration of the hydrogen ion, (H^+). Taken together, these results support the hypothesis that elevated atmospheric CO_2 can be shown to increase soil pCO_2 and concentrations of DIC. In addition to increasing concentrations of carbon in soil solution, Karberg et al. (2004) demonstrated that elevated atmospheric CO_2 alters system inorganic carbonate chemistry by increasing alkalinity, with the potential to increase weathering rates of primary minerals. These recent results demonstrate a direct link between altered photosynthesis, root growth, soil respiration, the biochemical feedbacks in the soil that can control rates of mineral weathering, and perhaps even an increase in the global DIC delivery to the oceans.

7.4   SUMMARY: CASCADING CONSEQUENCES OF ALTERED PRIMARY PRODUCTIVITY

It seems clear that the physiological response of plant root systems to a changing environment will cascade through terrestrial ecosystems to alter higher trophic levels (decomposers), which further regulate the flow of energy and nutrients in terrestrial ecosystems. A major uncertainty in our ability to predict ecosystem response to changing atmospheric chemistry is the extent to which fine root growth and tissue chemistry will be altered as the Earth's atmosphere changes. For example, we still do not understand how atmospheric N deposition will alter root growth, root mortality, and root tissue chemistry. Part of this problem is likely related to the issue of exactly how we define the pool of carbon and nutrients in fine roots. We also believe that the inherent physiological responses of plants to altered atmospheric CO_2 and soil N availability will depend on their life histories, and that mechanistic responses will not always follow the same path. Together, the underlying physiological responses of primary producers and microbial decomposers regulate the cycling of C and N in terrestrial ecosystems. In many regions of Earth, atmospheric CO_2 will increase in concert with increasing N deposition, but far too little attention has been given to the interaction of these changes in atmospheric chemistry. The impact of the interaction between elevated atmospheric CO_2 and O_3 on soil C storage clearly demonstrates why it is important to examine the interacting effects of our changing atmosphere. If we deliberately set out to understand how variable plant root and microbial physiology are to changes in atmospheric CO_2 and soil N, it should be possible to build a deeper understanding of the fundamental processes controlling ecosystem response to climate change.

ACKNOWLEDGEMENTS

This synthesis was supported by the US Department of Energy's Office of Science (PER: Program for Ecosystem Research), the USDA Forest Service (Northern Global Change Program and North Central Research Station), the National Science Foundation (DEB, DBI/MRI).

REFERENCES

Andrews, J.A., and W.H. Schlesinger. 2001. Soil CO_2 dynamics, acidification, and chemical weathering in a temperate forest with experimental CO_2 enrichment. *Global Biogeochemical Cycles* 15(1):149–162.

Berg, A., and S.A. Banwart. 2000. Carbon dioxide mediated dissolution of Ca-feldspar: Implications for silicate weathering. *Chemical Geology* 163:25–42.

Berner, R.A. 1997. The rise of plants and their effect on weathering and atmospheric CO_2. *Science* 276:544–546.

Berthelin, J. 1988. Microbial weathering processes in natural environments. In: Lerman, A., and Meybeck, M. (eds) *Physical and Chemical Weathering in Geochemical Cycles*. Kluwer Academic Publ., Dordrecht, The Netherlands, pp. 33–59.

Bormann, B.T., D. Wang, F.H. Bormann, G. Benoit, R. April, and R. Snyder. 1998. Rapid, plant-induced weathering in an aggrading experimental ecosystem. *Biogeochemistry* 43:129–155.

Bowden, R.D., E. Davidson, C. Arabia, K. Savage, and P. Steudler. 2004. Chronic nitrogen additions reduce total soil respiration and microbial respiration in temperate forest soils at the Harvard Forest. *Forest Ecology and Management* 196:43–56.

Bredemeier, M., K. Blanck, Y.J. Xu, A. Tietema, W.A. Boxman, B. Emmett, F. Moldan, P. Gundersen, P. Schleppi, and R.F. Wright. 1998. Input-output budgets at the NITREX sites. *Forest Ecology and Management* 101:57–64.

Bryant, J.P., F.S. Chapin, III, and D.R. Klein. 1983. Carbon/nutrient balance of boreal plants in relation to herbivory. *Oikos* 40:357–368.

Burton, A.J., K.S. Pregitzer, G.P. Zogg, and D.R. Zak. 1996. Latitudinal variation in sugar maple fine root respiration. *Canadian Journal of Forest Research* 26:1761–1768.

Burton, A.J., K.S. Pregitzer, and R.L. Hendrick. 2000. Relationships between fine root dynamics and nitrogen availability in Michigan northern hardwood forests. *Oecologia* 125:389–399.

Burton, A.J., K.S. Pregitzer, R.W. Ruess, R.L. Hendrick, and M.F. Allen. 2002. Root respiration in North American forests: Effects of nitrogen concentration and temperature across biomes. *Oecologia* 131:559–568.

Burton, A.J., K.S. Pregitzer, J.N. Crawford, G.P. Zogg, and D.R. Zak. 2004. Simulated chronic NO_3^- deposition reduces soil respiration in northern hardwood forests. *Global Change Biology* 10:1080–1091.

Carreiro, M.M., R.L. Sinsabaugh, D.A. Repert, and D.F. Parkhurst. 2000. Microbial enzyme shifts explain litter decay responses to simulated nitrogen deposition. *Ecology* 81:2359–2365.

Ceulemans, R., and M. Mousseau. 1994. Effects of elevated atmospheric CO_2 on woody plants. *New Phytologist* 127:425–446.

Chameides, W.L., P.S. Kasibhatla, J. Yienger, and H.I. Levy. 1994. Growth of continental-scale metro-agro-plexes, regional ozone pollution, and world food production. *Science* 264:74–77.

Clark, I.D., and P. Fritz. 1997. *Environmental Isotopes in Hydrogeology*, CRC Press, New York.

Coleman, M.D., R.E. Dickson, J.G. Isebrands, and D.F. Karnosky. 1995. Carbon allocation and partitioning in aspen clones varying in sensitivity to tropospheric ozone. *Tree Physiology* 15:593–604.

Collins, P.J., and A.D.W. Dobson. 1997. Regulation of Laccase gene transcription in *Tramates versicolor*. *Applied and Environmental Microbiology* 63(9):3444–3450.

Curtis, P.S. 1996. A meta-analysis of leaf gas exchange and nitrogen in trees grown under elevated carbon dioxide. *Plant Cell Environment* 19:127–137.

Curtis, P.S., and X. Wang. 1998. A meta-analysis of elevated CO_2 effects on woody plant mass, form and physiology. *Oecologia* 113:299–313.

DeLucia, E.H., J.G. Hamilton, S.L. Naidu, R.B. Thomas, J.A. Andrews, A. Finiz, M. Lavine, R. Matamala, J.E. Mohan, G.R. Hendry, and W.H. Schlesinger. 1999. Net primary production of a forest ecosystem with experimental CO_2 enrichment. *Science* 284:1177–1179.

Dickson, R.E., K.F. Lewin, J.G. Isebrands, M.D. Coleman, W.E. Heilman, D.F. Riemenschneider, J. Sôber, G.E. Host, D.R. Zak, G.R. Hendrey, K.S. Pregitzer, and D.F. Karnosky. 2000. Forest Atmosphere Carbon Transfer and Storage (FACTS-II) – The Aspen Free-air CO_2 and O_3 Enrichment (FACE) project: An overview. USDA Technical Report NC-214, Washington, DC.

Drever, J.I. 1994. The effect of land plants on weathering rates of silicate minerals. *Geochimica et Cosmochimica Acta* 58(10):2325–2332.

Eissenstat, D.M. 1992. Cost and benefits of constructing roots of small diameter. *Journal of Plant Nutrition* 15:763–782.

Eissenstat, D.M., and R.D. Yanai. 1997. The ecology of root lifespan. *Advances in Ecological Research* 27:1–60.

Fahey, T.J., and J.W. Hughes. 1994. Fine root dynamics in a northern hardwood forest ecosystem, Hubbard Brook Experimental Forest. NH. *Journal of Ecology* 82:533–548.

Field, C., and H.A. Mooney. 1986. The photosynthesis-nitrogen relationship in wild plants. In: Givninish, T. (ed.) *On the Economy of Plant form and Function*. Cambridge University Press, London, UK, pp. 24–44.

Findlay, S., M. Carreiro, V. Krischik, and C.G. Jones. 1996. Effects of damage to living plants on leaf litter quality. *Ecological Applications* 6:269–275.

Fog, K. 1988. The effect of added nitrogen on the rate of decomposition of organic matter. *Biological Reviews* 63:433–462.

Galloway, J.N. 1995. Acid deposition: Perspectives in time and space. *Water, Air and Soil Pollution* 85:15–24.

Gill, R.A., and R. Jackson. 2000. Global patterns of root turnover for terrestrial ecosystems. *New Phytologist* 147:13–31.

Guo, D.L., R.J. Mitchell, and J.J. Hendricks. 2004. Fine root branch orders respond differentially to carbon source-sink manipulations in a longleaf pine forest. *Oecologia* 140:450–457.

Hamilton, J.G., A.R. Zangerl, E.H. DeLucia, and M.R. Berenbaum. 2001. The carbon-nutrient balance hypothesis: Its rise and fall. *Ecology Letters* 4:86–95.

Hammel, K.E. 1997. Fungal degradation of lignin. In: Cadish, G., and Giller, K.E. (eds) *Driven by Nature: Plant Litter Quality and Decomposition*. International Symposium, Wye, England, UK, September 17–20, 1995. CAB International, Wallingford, England, UK, pp. 33–45.

Hanson, P.J., N.T. Edwards, C.T. Garten, Jr., and J.A. Andrews. 2000. Separating root and soil microbial contributions to soil respiration: A review of methods and observations. *Biogeochemistry* 48(1):115–146.

Haynes, B.E., and S.T. Gower. 1995. Belowground carbon allocation in unfertilized and fertilized plantations in northern Wisconsin. *Tree Physiology* 15:317–325.

Hendrick, R.L., and K.S. Pregitzer. 1992. The demography of fine roots in a northern hardwood forest. *Ecology* 73(3):1094–1104.

Herms, D.A., and W.J. Mattson. 1992. The dilemma of plants: To growth or defend. *Quarterly Review of Biology* 67:283–335.

Homann, P.S., B.A. Caldwell, H.N. Chappell, P. Sollins, and C.W. Swanston. 2001. Douglas-fir soil C and N properties a decade after termination of urea fertilization. *Canadian Journal of Forest Research* 31:2225–2236.

Kane, E.S., K.S. Pregitzer, and A.J. Burton. 2003. Soil respiration along environmental gradients in Olympic National Park. *Ecosystems* 6:326–335.

Karberg, N.J., K.S. Pregitzer, J.S. King, A.L. Friend, and J.R. Wood. 2004. Soil pCO_2 and dissolved inorganic carbonate chemistry under elevated CO_2 and O_3. *Oecologia* 142:296–306.

King, J.S., K.S. Pregitzer, D.R. Zak, J. Sober, J.G. Isebrands, R.E. Dickson, G.R. Hendry, and D.F. Karnosky. 2001a. Fine root biomass and fluxes of soil carbon in young stands of paper birch and trembling aspen as affected by elevated CO_2 and tropospheric O_3. *Oecologia* 128:237–250.

King, J.S., K.S. Pregitzer, D.R. Zak, M.E. Kubiske, and W.E. Holmes. 2001b. Correlation of foliage and litter chemistry of sugar maple, *Acer saccharum*, as affected by elevated CO_2 and varying N availability, and effects on decomposition. *Oikos* 94:403–416.

King, J.S., T.J. Albaugh, H.L. Allen, M. Buford, B.R. Strain, and P. Dougherty. 2002. Seasonal dynamics of fine roots relative to foliage and stem growth in loblolly pine (*Pinus taeda* L.) as affected by water and nutrient availability. *New Phytologist* 154:389–398.

King, J.S., K.S. Pregitzer, D.R. Zak, W.E. Holmes, and K. Schmidt. 2005. Fine root chemistry and decomposition in model communities of north-temperate tree species show little response to elevated CO_2 and varying soil resource availability. *Oecologia* 146:318–328.

Koricheva, J., S. Larsson, E. Haukioja, and M. Keinanen. 1998. Regulation of woody plant secondary metabolism by resource availability: Hypothesis testing by means of meta-analysis. *Oikos* 83:212–226.

Lackner, K.S. 2002. Carbonate chemistry for sequestering fossil carbon. *Annual Review of Energy and the Environment* 27:193–232.

Larson, J.L., D.R. Zak, and R.L. Sinsabaugh. 2002. Microbial activity beneath temperate trees growing under elevated CO_2 and O_3. *Soil Science Society of America* 66:1848–1856.

Leake, J.R., D.P. Donnelly, E.M. Saunders, L. Boddy, and D.J. Read. 2001. Rates and quantities of carbon flux to ectomycorrhizal mycelium following ^{14}C pulse labeling of *Pinus sylvestris* seedlings: Effects of litter patches and interaction with a wood-decomposer fungus. *Tree Physiology* 21:71–82.

Leavitt, S.W., R.F. Follett, and E.A. Paul. 1996. Estimation of slow- and fast-cycling soil organic carbon pools from 6N HCl hydrolysis. *Radiocarbon* 38:231–239.

Loomis, W.E. 1932. Growth-differentiation balance vs. Carbohydrate-nitrogen ratio. *Proceedings of the American Society of Horticultural Science* 29:240–245.

Lorio, P.L. 1986. Growth-differentiation balance: A basis for understanding southern pine beetle–tree interactions. *Forest Ecology and Management* 14:259–273.

Loya, W.M., K.S. Pregitzer, N.J. Karberg, J.S. King, and C.P. Giardina. 2003. Reduction of soil carbon formation by tropospheric ozone under elevated carbon dioxide levels. *Nature* 425:705–707.

MacDonald, N.W., A.J. Burton, M.F. Jurgensen, J.W. McLaughlin, and G.D. Mroz. 1991. Variation in forest soil properties along a great lakes air pollution gradient. *Soil Science Society of America Journal* 55:1709–1715.

Matamala, R., M.A. Gonzàlez-Meler, J.D. Jastrow, R.J. Norby, and W.H. Schlesinger. 2003. Impacts of fine root turnover on forest NPP and soil C sequestration potential. *Science* 21:1385–1387.

Mooney, H.A., K. Fichtner, and E.-D. Schulze. 1995. Growth, photosynthesis, and storage of carbohydrates and nitrogen in *Plaseolus lunatus* in relation to resource availability. *Oecologia* 104:17–23.

Nadelhoffer, K.J., J.D. Aber, and J.M. Melillo. 1985. Fine roots, net primary productivity, and soil nitrogen availability: A new hypothesis. *Ecology* 66:1377–1390.

Nadelhoffer, K.J., and J.W. Raich. 1992. Fine root production estimates and belowground carbon allocation in forest ecosystems. *Ecology* 73:1139–1147.

Nohrstedt, H.-Õ., and G. Börjesson. 1998. Respiration in a forest soil 27 years after fertilization with different doses of urea. *Silva Fennica* 32:383–388.

Norby, R.J., J. Ledford, C.D. Reilly, N.E. Miller, and E.G. O'Neill. 2004. Fine-root production dominates response of a deciduous forest to atmospheric CO_2 enrichment. *PNAS* 101:9689–9693.

Oren, R., D.S. Ellsworth, K.H. Johnson, N. Phillips, B.E. Ewers, C. Maier, K.V.R. Schäfer, H. McCarthy, G. Hendrey, S.G. McNulty, and G.G. Katul. 2001. Soil fertility limits carbon sequestration by a forest ecosystem in a CO_2-enriched atmosphere. *Nature* 411:469–472.

Pankow, J.F. 1991. *Aquatic Chemistry Concepts*. Lewis Publishers, Chelsea, MI.

Paul, E.A., R.F. Follett, S.W. Leavitt, A. Halvorson, G.A. Peterson, and D.J. Lyon. 1997. Radiocarbon dating for determination of soil organic matter pool sizes and dynamics. *Soil Science Society of America Journal* 61:1058–1067.

Percy, K.E., C.S. Awmack, R.L. Lindroth, M.E. Kubiske, B.J. Kopper, J.G. Isebrands, K.S. Pregitzer, G.R. Hendrey, R.E. Dickson, D.R. Zak, F. Oksanen, J. Sober, R. Harrington, and D.F. Karnosky. 2002. Altered performance of forest pests under atmospheres enriched by CO_2 and O_3. *Nature* 420:403–407.

Phillips, R.L., D.R. Zak, and W.E. Holmes. 2002. Microbial community composition and function beneath temperate trees exposed to elevated atmospheric CO_2 and O_3. *Oecologia* 131:236–244.

Pregitzer, K.S. 2002. The fine roots of trees – a new perspective. *New Phytologist* 154:267–273.

Pregitzer, K.S., D.R. Zak, P.S. Curtis, M.E. Kubiske, J.A. Teeri, and C.S. Vogel. 1995. Atmospheric CO_2, soil nitrogen and turnover of fine roots. *New Phytologist* 129:579–585.

Pregitzer, K.S., M.J. Laskowski, A.J. Burton, V.C. Lessard, and D.R. Zak. 1998. Variation in sugar maple root respiration with root diameter and soil depth. *Tree Physiology* 18:665–670.

Pregitzer, K.S., J.A. DeForest, A.J. Burton, M.F. Allen, R.W. Ruess, and R.L. Hendrick. 2002. Fine root architecture of nine North American trees. *Ecological Monographs* 72:293–309.

Reich, P.B., M.B. Walters, and D.S. Ellsworth. 1997. From tropics to tundra: Global convergence in plant functioning. *Proceedings of the National Academy of Sciences USA* 94:13730–13734.

Rogers, H.H., G.B. Runion, and S.V. Krupa. 1994. Plant responses to atmospheric CO_2 enrichment with emphasis on roots and the rhizosphere. *Environmental Pollution* 83:155–189.

Ruess, R.W., R.L. Hendrick, A.J. Burton, K.S. Pregitzer, B. Sveinbjornssön, M.F. Allen, and G.E. Maurer. 2003. Coupling fine root dynamics with ecosystem carbon cycling in black spruce forests of interior Alaska. *Ecological Monographs* 73:643–662.

Ryan, M.G. 1991. Effects of climate change on plant respiration. *Ecological Applications* 1:157–167.

Ryan, M.G., R.M. Hubbard, S. Pongracic, R.J. Raison, and R.E. McMurtrie. 1996. Foliage, fine root, woody-tissue and stand respiration in Pinus radiate in relation to nitrogen status. *Tree Physiology* 16:333–343.

Ryan, M.G., D. Binkley, and J.H. Fownes. 1997. Age-related decline in forest productivity: Pattern and process. *Advances in Ecological Research* 27:213–262.

Sinsabaugh, R.L., M.M. Carreiro, and D.A. Repert. 2002. Allocation of extracellular enzyme activity in relation to litter composition, N deposition and mass loss. *Biogeochemistry* 60:1–24.

Six, J., R.T. Conant, E.A. Paul, and K. Paustian. 2002. Stabilization mechanisms of soil organic matter: Implications for C-saturation of soils. *Plant and Soil* 241:155–176.

Söderström, B., E. Bååth, and B. Lundgren. 1983. Decrease in soil microbial activity and biomass owing to nitrogen amendments. *Canadian Journal of Microbiology* 29:1500–1506.

Stevenson, F.J. 1994. Humus chemistry, genesis, composition, reactions, 2nd edn. John Wiley & Sons, Inc. New York.

Tierney, G., and T. Fahey. 2001. Evaluating minirhizotron estimates of fine root longevity and production in the forest floor of a temperate broadleaf forest. *Plant Soil* 229:167–176.

Tingey, D.T., D.L. Phillips, and M.G. Johnson. 2000. Elevated CO_2 and conifer roots: Effects on growth, life span and turnover. *New Phytologist* 147:87–103.

Trumbore, S.E., and J.B. Gaudinski. 2003. The secret lives of roots. *Science* 302:1344–1345.

Vitousek, P.M., J.D. Aber, R.W. Howarth, G.E. Likens, P.A. Matson, D.W. Schindler, W.H. Schlesinger, and D.G. Tilman. 1997. Human alteration of the global nitrogen cycle: Sources and consequences. *Ecological Applications* 7:737–750.

Vogt, K.A., D.J. Vogt, S.T. Gower, and C.C. Grier. 1990. Carbon and nitrogen interactions for forest ecosystems. In: Persson, H. (ed.) *Above and Belowground Interactions in Forest Trees in Acidified Soils.* Commission of the European Communities, Belgium, pp. 203–235.

Waldrop, M.P., D.R. Zak, and R.L. Sinsabaugh. 2004. Microbial community response to nitrogen deposition in northern forest ecosystems. *Soil Biology and Biochemistry* 36:1443–1451.

Wells, C.E., and D.M. Eissenstat. 2001. Marked differences in survivorship among apple roots of different diameters. *Ecology* 82:882–892.

Withington, J.M., A.D. Elkin, B. Bulaj, J. Olesiński, J.N. Tracy, T.J. Bouma, J. Oleksyn, L.J. Anderson, J. Modrzyński, P.B. Reich, and D.M. Eissenstat. 2003. The impact of material used for minithizotron tubes for root research. *New Phytologist* 160:533–544.

Zak, D.R., and K.S. Pregitzer. 1990. Spatial and temporal variability of nitrogen cycling in northern lower Michigan. *Forest Science* 36:367–380.

Zak, D.R., and K.S. Pregitzer. 1998. Integration of ecophysiological and biogeochemical approaches to ecosystem dynamics. In: Pace, M., and Groffman, P. (eds) *Successes, Limitations and Frontiers in Ecosystem Science.* Springer-Verlag, New York, pp. 372–403.

Zak, D.R., K.S. Pregitzer, J.S. King, and W.E. Holmes. 2000a. Elevated atmospheric CO_2, fine roots and the response of soil microorganisms: A review and hypothesis. *New Phytologist* 147:201–222.

Zak, D.R., K.S. Pregitzer, P.S. Curtis, C.S. Vogel, W.E. Holmes, and J. Lussenhop. 2000b. Atmospheric CO_2, soil N availability, and the allocation of biomass and nitrogen in Populus tremuloide. *Ecological Applications* 10:34–46.

CHAPTER 8

The Rhizosphere and Soil Formation

Daniel deB. Richter, Neung-Hwan Oh, Ryan Fimmen, and Jason Jackson

> There are not many differences in mental habit more significant than that between thinking in discrete, well defined class concepts and that of thinking in terms of continuity, of infinitely delicate shading of everything into something else, of the overlapping of essences, so that the whole notion of species comes to seem an artifact of thought, not truly applicable to fluency, the so to say universal overlapping of the real world.
>
> A.O. Lovejoy (1936)

8.1 INTRODUCTION

By most accounts, the rhizosphere is narrowly conceived in space and time. Since first described by Hiltner (1904), the rhizosphere is taken as the soil volume that interacts directly and immediately with living plant roots, the near-root environment nanometers to centimeters in radial distance from the root surface. As intimate interface between roots and the mineral world, rhizospheres are remarkable environments, and have ecological feedbacks, chemical interactions, and inter-organism communication as complex as any in the aboveground world. There are excellent reasons that the rhizosphere concept has been narrowly focused in its first 100 years of use, and that it is distinguished from the bulk soil, that is the soil not in direct and immediate interaction with active roots.

Yet, over pedogenic time, all of the soil's A and B horizons are greatly influenced by plant roots. In fact, this chapter is written to advance the idea that rhizospheres typically affect, even transform, a large soil environment, that is all of the so-called "bulk soil." Although not often appreciated, rhizosphere processes stimulate mineral weathering and direct the ultimate formation of soils. While the narrow definition of the rhizosphere has helped emphasize that actively growing roots create unique and special environments with great consequence for plants and microbes, the rhizosphere also has a wide range of significant effects on soil formation and biogeochemistry. The rhizosphere is the critical interface between biota and geologic environment, the locale where roots exert intense physical pressures on surrounding soils, the chemical environment where biogeochemical reactions interact with minerals, and the special habitat for a wide assemblage of well-adapted microbes (see Chapters 1, 3, and 4). Rhizospheres are thus fundamentally important to soil formation, including the formation of the earth's most extremely weathered soils the Ultisols and Oxisols (Richter and Babbar 1991).

This chapter examines rhizospheres and some of their broad biological, physical, and chemical effects on soil formation. In organization, the chapter opens with a discussion of general concepts: of the rhizosphere vs bulk soil dichotomy, of rhizospheres as microsites within soil profiles, and of soil formation including the formation of advanced weathering-stage soils. Subsequently, we evaluate a number of the physical and chemical effects of rooting on the soil. Throughout, the biota's physical and chemical interactions with soils are seen to be concentrated in the rhizosphere, and over time these interactions transform soils across a wide range of spatial scales, from individual mineral grains to entire soil horizons and profiles. We conclude that rhizosphere processes are instrumental to soil formation including even the earth's most advanced weathering-stage soils. Throughout this chapter we use data from our long-studied research ecosystem at the Calhoun Experimental Forest in the South Carolina Piedmont and the Duke University Forest (Richter and Markewitz 2001) to support our perspectives of the rhizosphere.

8.2   A REVIEW OF CONCEPTS

RHIZOSPHERE VS BULK SOIL

Plant roots, that is rhizospheres, are networks within the bulk soil, biological hotspots where respiration, gas exchange, nutrient and moisture use, and localized supplies of organic matter are most concentrated (Curl and Truelove 1986). In contrast, the bulk soil is a more oligotrophic environment, especially with respect to supply of root-derived organic matter. More than anything, reactive organic reductants and microbial activity are concentrated near roots compared with the soil as a whole.

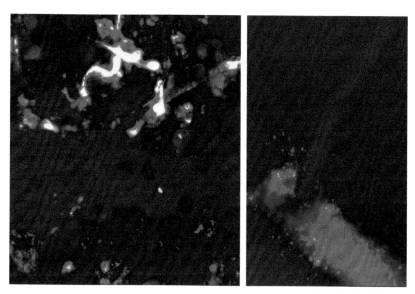

PLATE 1 *Avena barbata* (slender wild oat) roots growing through soil. On the left, magnification is 100×. Plant root and root hairs autofluoresce blue, and soil aggregates infested with bacteria are visible in black at the bottom. Rhizosphere was inoculated with bacteria marked with a constitutively expressing dsRed protein, so all introduced bacteria are visible as red dots. On the right, magnification is 1000×, and bacteria can be seen colonizing the nook between the root and the emerging root hair. Photos by K. DeAngelis. See Figure 1.1, p. 2.

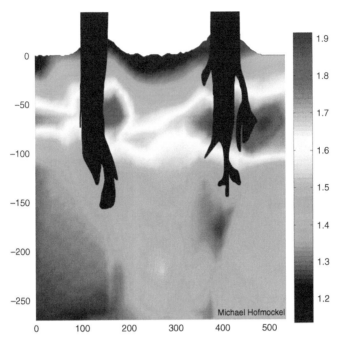

PLATE 2 Bulk density of soil surrounding two 70-year-old loblolly pine trees. Bulk density in g/cm^3 (scaled in color in key to right of figure). Depth and horizontal distances are in cm (on y- and x- axis respectively). Bulk densities were obtained with conventional slide hammer for 180 samples on the face of the excavation. Isolines of densities were obtained using Matlab's interpolation via a shading function ("INTERP"). See Figure 8.5, p. 189.

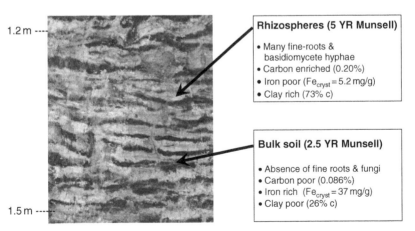

Rhizospheres (5 YR Munsell)

- Many fine-roots & basidiomycete hyphae
- Carbon enriched (0.20%)
- Iron poor (Fe$_{cryst}$ = 5.2 mg/g)
- Clay rich (73% c)

Bulk soil (2.5 YR Munsell)

- Absence of fine roots & fungi
- Carbon poor (0.086%)
- Iron rich (Fe$_{cryst}$ = 37 mg/g)
- Clay poor (26% c)

PLATE 3 Pronounced rhizosphere-initiated redoximorphic features that demonstrate effects of Fe-redox cycling in B horizons at Calhoun Experimental Forest, SC. The close-up photo is of a soil excavation at 1.2–1.5 m depth. See Figure 8.6, p. 193.

By convention (and as an example of Lovejoy's (1936) class concept), the rhizosphere has been characterized as having three components (Clark 1949):

1. *rhizoplane*, the immediate surface of the root,
2. *rhizosphere*, the soil volume surrounding the rhizoplane that is immediately affected by root activity, and
3. *bulk soil*, the soil not directly affected by living roots.

This tripartite construct helps emphasize the special nature of the rhizosphere, but we suggest that it overemphasizes a dichotomy between the rhizosphere and bulk soils. Although the concept of rhizosphere has hardly been monolithic (e.g., Rovira and Davey 1974), a neat division of rhizosphere and bulk soil is difficult to align with our developing understanding of root systems and their effects on soil. High-powered microscopy (e.g., scanning electron microscopy) demonstrates that the rhizoplane is far from a planer surface, and a variety of investigations indicate that the radial influence of the rhizosphere is very ill-defined and that it ranges widely in spatial scale (e.g., Rovira and Davey 1974). Root systems are symbiotic systems in which cells of plants, fungi, and bacteria are intimately associated, both structurally and functionally, so much so that it is difficult to isolate what is plant from what is microbe. The fact that fungi and bacteria colonize root tissues in "endorhizospheres" suggests that concepts of continuity rather than those of class may be in order for how we think of rhizospheres and soil. In place of class concepts of rhizoplane, rhizosphere, and bulk soil, a continuum seems much more pertinent between the following:

- *root–microbe system*, which includes all cells of plant roots, mycorrhizal fungi, and closely associated non-mycorrhizal fungi and bacteria;
- *rhizosphere surrounding these cells*, a volume which is immediately affected by the functioning of the root–microbe system and depends on chemical reaction, chemical element, microorganism, and soil type; and
- *bulk soil*, the soil not immediately affected by the active functioning of roots, but which may be transformed by rhizospheres over pedogenic time.

Much rhizosphere research, however, including our own, relies heavily on a dichotomous contrast of characteristics or processes of the rhizosphere with those of the bulk soil. Whether the variable of interest is microorganism numbers, organic compounds, biological or chemical reactions, or communication-signaling, "rhizosphere effects" are frequently indexed by R/S ratios, that is the ratio of an attribute in the rhizosphere to that in bulk soil (Katznelson 1946). For many soils, R/S ratios for microorganism numbers range from 5 to 20 to

TABLE 8.1   Chemistry and Microbial Properties of Bulk Soil (Conventional 6 cm dia Core Samples) in Four Soil Horizons, and in Rhizospheres (<2 mm Distance from Roots) Sampled at 2–3 m Depth in the Pine-Forest Soil of the Calhoun Experimental Forest. Soil Microbial Data Courtesy of Dr Elaine Ingham, Oregon State University, Corvallis

Soil material	Soil depth (m)	Total carbon (%)	Total bacteria (cells g^{-1})	FDA*-active bacteria (cells g^{-1})	Total fungi (m g^{-1})	FDA-active fungi (m g^{-1})
Oe horizon	–	–	1.97×10^8	32.9×10^6	59160	906
A horizon	0–0.075	0.70	1.44×10^8	23.8×10^6	18140	653
BE horizon	0.6–1.0	0.24	1.59×10^8	1.47×10^6	294	5.5
B horizon	2.0–3.0	0.073	1.23×10^8	0**	0**	0**
Rhizosphere soil in B	2.0–3.0	0.42	3.17×10^8	3.54×10^6	1467	65.8

* Fluorescein diacetate stain.

** Detectable concentrations for FDA-active bacteria, total fungi, and FDA-active fungi are $<4 \times 10^3$ units g^{-1}, <0.3 cm g^{-1}, <0.3 cm g^{-1}, respectively.

even > 100 (Richter and Markewitz 2001; Anderson *et al.* 2002). Especially deep in the soil, active bacteria and fungi may be prolific in the rhizosphere but approach limits of detection in the surrounding soil (Table 8.1).

Approaches to the rhizosphere based on R/S ratios have been instructive in emphasizing the biological and chemical activity of the habitat of the near-root environment. Unfortunately, R/S ratios suggest a lack of interaction between the rhizosphere and the bulk soils. This is important as several rhizosphere processes significantly interact with bulk soils over pedogenic time.

By broadening perspectives of the rhizosphere, we by no means oppose traditional concepts of the rhizosphere, although we do wish to promote an appreciation for how biological, chemical, and physical activity near roots have profound effects on the whole soil, especially when integrated over pedogenic time. In fact, interactions between the rhizosphere and the whole soil make research on these issues some of the most exciting in all of soil science, biogeochemistry, and ecosystem ecology.

RHIZOSPHERES AS MICROSITES WITHIN SOIL PROFILES

Soil scientists and ecologists have long divided the soil profile into an upper "solum" and the lower "parent material," in part due to the physical and chemical effects of rooting. The solum is taken to be the O, A, and B horizons, the parent material the C horizon. Rhizosphere densities are much higher in the A and B horizons, but in many soils rhizospheres extend well into the C horizon. The upper soil system – that is the O, A, and B horizons – is characterized by intense biological activity, a variety of ecological processes, and extensive and thorough rooting (Table 8.1). Roots and associated microorganisms affect much of the physics and chemistry of the upper soil system (Chadwick *et al.* 1990; Brimhall *et al.* 1991; Richter and Markewitz 1995).

With increasing soil depth, concentrations typically diminish of roots, active microbes, organic matter, and bioavailable nutrients. In the soil's lower system, deep within B and throughout C horizons, the near-root environment is nothing less than an oasis of resources compared with the surrounding subsoil. In some respects, rhizospheres in the lower soil system have more in common with the A horizons than they do with the B and C horizons that surround them (Table 8.1). The R/S ratios for biologic and chemical properties may well increase with increasing soil depth (Figure 8.1), a pattern indicative

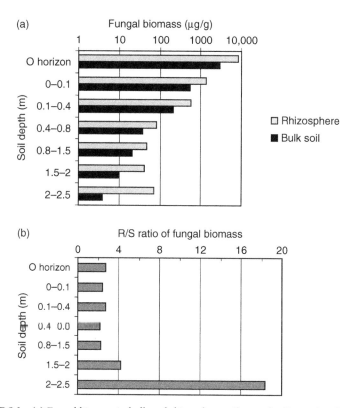

FIGURE 8.1   (a) Fungal biomass in bulk and rhizosphere soil at an Appling soil at the Calhoun Experimental Forest, South Carolina. (b) The conventional R/S ratio for fungal biomass as a function of soil depth. Soil supported a 47-year-old loblolly pine (*Pinus teada*) forest. Fungal hyphae in the rhizosphere soil are illustrated in Figure 8.3. Soil fungal data courtesy of Dr Elaine Ingham, Oregon State University, Corvallis.

of the functioning and structure of rhizospheres in lower soil horizons. For example, in our research site in South Carolina, fungal biomass in bulk soil decreases steadily by three orders of magnitude from the soil surface to 2.5 m depth, whereas fungal biomass in rhizospheres remains relatively constant between depths of 0.4 and 2.5 m.

SOIL FORMATION

Because soils are open thermodynamic systems, soils experience a remarkable set of transformations over time, as energy, chemical elements, and water are processed. Over time, primary minerals are weathered and lost. Although new secondary minerals may be formed during soil development, the soil's primary minerals are decomposed and its acid-neutralizing capacity gradually consumed. If the soil's landform is geomorphically stable, weathering of soils may proceed through a full sequence of weathering as illustrated by Jackson and Sherman (1953) in Table 8.2. Over pedogenic time, weathering consumes even large pools of primary minerals and advanced weathering-stage soils will be formed if hydrologic removals of solutes outpace renewals that can come from weatherable minerals or atmospheric deposition. Our interest in this chapter is in exploring how rhizospheres are involved in the advancement of weathering and soil formation, even including the formation of the earth's most weathered soils, the Ultisols and Oxisols (Richter and Babbar 1991).

TABLE 8.2   Soils are Open Thermodynamic Systems, and Over Time are Transformed, as Energy, Chemical Elements, and Water are Processed. Three General Weathering Stages of Soil Mineral Weathering were Used by Jackson and Sherman (1953) to Illustrate Soil Formation. The Implications of this System Change for Common Soil Minerals and Soil Orders is Illustrated in the Table. This Paper Illustrates the Fundamental Importance of Rhizospheres to the Weathering of Minerals and Formation of Soils

| Attribute | Jackson-Sherman (1953) soil weathering stage and soil formation | | |
	Early	Intermediate	Advanced
Soil Taxonomy orders (Soil Survey Staff 1998)	Entisol Andisol	Inceptisol Mollisol Alfisol	Ultisol Oxisol
Common soil minerals	Gypsum Calcite Olivine Biotite Feldspar	Feldspar Muscovite Vermiculite Smectite	Kaolinite Gibbsite Fe oxide/hydroxides

In humid temperate zones and the tropics, geomorphically stable surfaces can develop enormously deep profiles, sometimes >20 m deep above unweathered bedrock. It is not uncommon that soil weathering exhausts all primary minerals and a number of chemical elements throughout these depths (Figure 8.2). Not atypical in advanced weathering-stage soils is an upper 1–3 m of O, A, and B horizons, below which is the C horizon of highly variable depth, all of which is acidic, extremely low in base cations and phosphorus, and depauperate in primary minerals. Since the original starting materials have been completely transformed by weathering, these soils are composed of only the most insoluble chemical elements and recalcitrant minerals. Only a few chemical elements, such as Zr and Ti, are insoluble enough to resist transportation from weathering environments. It is easy to underestimate the extreme state of weathering exhibited by such soils, and we suggest easy to underestimate the weathering as affected by rhizosphere processes.

Several calculations help emphasize the extreme state of weathering represented by such soils. In our long-term research site at the Calhoun Experimental Forest in the Piedmont of South Carolina, unweathered granite and gneiss underlies A, B, and C horizons in soil profiles that may total up to 25 m of unconsolidated material over actively weathering bedrock. The pH of ground samples of pulverized but unweathered bedrock is 7.9 in water, yet the pH of the soil sampled throughout at least the upper 8 m of A to C horizons

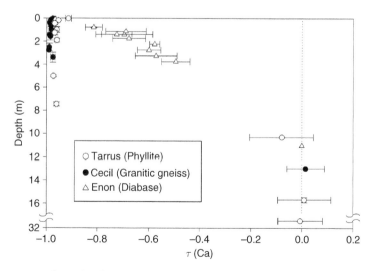

FIGURE 8.2 Calcium loss from three deep soil profiles (Tarrus and Cecil are Ultisols, Enon is an Alfisol). Tau expresses the estimate of the original Ca that has been lost during soil formation (e.g., −0.6 indicates that 60 percent of the Ca in the primary minerals has been lost to weathering).

ranges from 3.8 to 4.2 in $0.01 \, M \, CaCl_2$. Exchangeable acidity (with $1 \, M \, KCl$) totals about $4000 \, k \, molc \, ha^{-1}$ in this 8 m soil profile, an enormous quantity of acidity. Even more impressive, however, is the quantity of acid that has been consumed during weathering of granitic-gneiss into the kaolinite-dominated Ultisol. Transforming granitic-gneiss into 1 m of kaolinite is estimated to require (i.e., to consume) on the order of $100,000 \, k \, molc \, ha^{-1}$ of acid (Richter and Markewitz 1995, 2001). Weathering 10 m of granitic gneiss to kaolinite thus requires about $10^6 \, k \, molc \, ha^{-1}$. This extreme acidification raises questions about the sources and rates of acid inputs that have so thoroughly weathered these Ultisols, much less advanced weathering-stage soils overall. In the next section of our chapter, we examine how the rhizosphere is responsible for a considerable fraction of the weathering that over pedogenic time leads to such advanced weathering-stage soils.

8.3 RHIZOSPHERES: WHERE ECOSYSTEMS CONCENTRATE BIOLOGICAL INTERACTIONS WITH SOIL MINERALS

The extreme acidification and weathering state of Ultisols and Oxisols raise questions about the mechanisms by which these soils are transformed over time. Since rooting affects both physical and chemical weathering in soils and rocks, in this section, we examine some mechanical effects of rooting on the soil environment, and subsequently examine prominent sources of rhizosphere acidity that stimulate weathering and soil formation.

THE PHYSICAL ATTACK

Growing roots and their mycorrhizal hyphae follow pores and channels that are generally not less than their own diameters (Figure 8.3). As tree roots grow, they expand in volume radially, and exert enormous pressures on the surrounding soil by cylindrical expansion. Even relatively consolidated, unweathered rocks are susceptible to physical effects of roots. Rock wedging results when growing roots expand rocks' planes of weakness in joints or fractures. Over generations of trees, root growth and tree uprooting facilitate mechanical weathering of minerals in A and B horizons, accelerating chemical weathering by increasing minerals' surface area that is contacted by microbes, organic compounds, electrons, and protons.

The pressure of growing roots can be so great that roots can fracture and decompose minerals by exerting pressures on individual mineral grains or whole soils, that is across spatial scales that range from sub-micrometers to

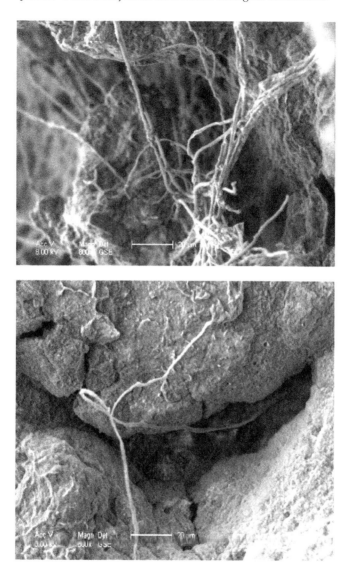

FIGURE 8.3   ESEM images (8.0 kV, 20 μm on horizontal scale) of rhizospheres at 1.5 m depth in Appling soil B horizon at the Calhoun Experimental Forest, SC. Rhizospheres are of basidiomycete hyphae of the genus, *Rhizopogon*.

many decimeters and even meters (Dexter 1987; Misra *et al.* 1987; April and Keller 1990; Richter *et al.* submitted).

In A horizons, growing roots can displace soil upward. Surrounding the root collars of large trees, for example, surface soils are uplifted considerably

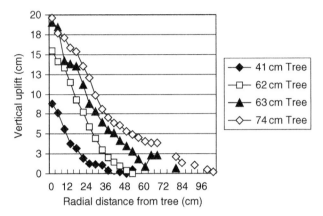

FIGURE 8.4 Soil microtopography surrounding four 70-year-old loblolly pine (*Pinus teada*) trees in the Duke Forest, North Carolina. Diameters are given for each of the four trees.

in the surrounding rhizosphere (Figure 8.4). Over time, the uprooting of trees especially during windstorms causes particle abrasion and mixing of the upper soil system, increasing the soil's surface area that is subject to chemical weathering.

In contrast to A horizons, root growth pressures cannot be relieved by upward displacement in B and C horizons. Pressures of growing roots are relieved by soil consolidation, as taproots establish anchorage by expanding radially in a process that has severe physical effects on individual soil particles, soil structure, and overall soil architecture. In the Duke Forest, bulk densities of B horizons adjacent to tap roots of 70-year-old trees exceeded $1.9 \, \mathrm{mg \, m^{-3}}$, a consequence of tap roots consolidating soil for up to 50 cm radial distance from the growing root (Figure 8.5). These rhizosphere effects no doubt cause severe abrasion and disintegration of individual soil particles, reduced porosity, hydraulic conductivity, and aeration, and greatly altered biogeochemical functioning. Such effects accumulate over time and may represent a significant, understudied process affecting biogeochemistry of forests. Such mechanical processes have impacted forested Ultisols and Oxisols on numerous occasions, given these soils' relatively great age.

THE CHEMICAL ATTACK

Rhizospheres not only physically alter soil minerals from individual grains to whole horizons, they chemically interact with soils at a wide range of spatial scales as well. Here, we focus on four rhizosphere processes that affect soil

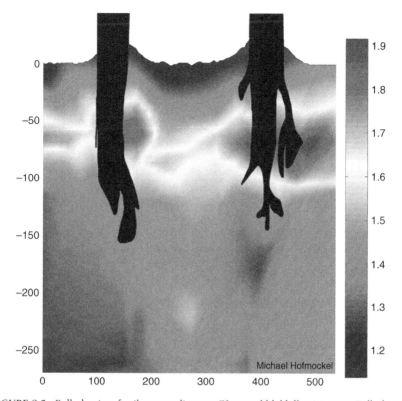

FIGURE 8.5   Bulk density of soil surrounding two 70-year-old loblolly pine trees. Bulk density in g/cm^3 (scaled in color in key to right of figure). Depth and horizontal distances are in cm (on y- and x-axis respectively). Bulk densities were obtained with conventional slide hammer for 180 samples on the face of the excavation. Isolines of densities were obtained using Matlab's interpolation via a shading function ("INTERP"). See Plate 2.

acidification and weathering, and thereby promote soil formation including that of advanced weathering-stage soils. The four rhizosphere processes that directly affect acid production include the following:

- Root nutrient-ion uptake;
- Organic acid production;
- Redox-reactions of metals;
- CO_2 production.

Remarkably, each of these sources of acidity result from the vegetative production and decomposition of photosynthetically derived organic matter. Although other biogenic acid systems can affect soil acidification and weathering dissolution (e.g., oxidation reactions involving nitrogen and sulfur), we focus on these four as widespread sources of protons across a wide range of rhizospheres.

Root Uptake of Nutrient Ions

A major source of soil acidity is derived from the uptake of nutrients by vegetation. Root uptake of nutrient cations and anions directly affects the soil's acid–base status because the physiological process of nutrient-ion uptake is electroneutral: that is the uptake of cations and anions is balanced by the release of H^+ and OH^-, respectively, into the rhizosphere. If roots take up more nutrients as cations than anions, plant accumulation of nutrients acidifies soil. A large collection of scientific literature describes soil acidification by "excess cation uptake" in terrestrial ecosystems, including cultivated field crops, aggrading secondary forests, and old-growth forests (Pierre *et al.* 1970; Sollins *et al.* 1980; Ulrich 1980; Driscoll and Likens 1982; van Breemen *et al.* 1982; Binkley and Richter 1987; Johnson and Lindberg 1992; Markewitz *et al.* 1998).

Plant species exert differential effects on soil acidity due in part to plant-nutrient uptake requirements. Many oak and hickory species (*Quercus* and *Carya* spp.) have calcium uptake that is two- to fivefold greater than many pines (*Pinus* spp.), and thus have much more potential to promote acidity throughout the rooting zone. Alban (1982) demonstrated this with comparisons of acidity in soils that supported tree species with a wide range of cation uptake, and we hypothesize that such species differences are expressed most greatly in rhizospheres. Richter (1986) estimated H^+ budgets of five forest stands and illustrated a manyfold variation in cation uptake and potential for soil acidification.

Within the rhizosphere, very low pH has been measured with plant systems having large net cation uptake (Lynch 1990). As much as two pH-unit depressions have been measured in rhizospheres compared with bulk soil, conditions that will affect not only cation exchange in the rhizosphere, but dissolution of weatherable minerals as well (April and Keller 1990).

Organic Acid Production

Organic acids play significant and varied roles in rhizosphere acidification and mineral weathering, contributing protons and serving as ligands that complex metals (Boyle and Voigt 1973; Duchaufour 1982; Brimhall *et al.* 1991; Qualls and Haines 1991; Buol *et al.* 1997). Organic acids can also promote redox reactions with electron-deficient metals (a rhizosphere-promoted process considered in the next section on redox cycling). A wide variety of organic compounds have acid functional groups, mainly carboxylic or phenolic, and these originate not only from products of decomposition and carbon oxidation but also as exudates from plant roots and associated microbes (Lapeyrie *et al.* 1987; Herbert and Bertsch 1995). Organic acids range widely in molecular weight from relatively small compounds such as oxalic and citric acids to much larger humic compounds with enormous numbers of carboxylic and phenolic functional groups.

Organic acids are weak acids with pKa values that range widely from as low as 3 (carboxylic) to as high as 9 (phenolic). Many carboxylic functional groups have a relatively low pKa, with oxalic, citric, malic, and formic acids, as representative low-molecular weight organic acids (Fox and Comerford 1990), all possessing pKa < 4.0. Such acids readily contribute protons to the soil system under a wide range of pH conditions. In addition to being a source of protons, many organic acids are effective ligands that complex metal cations such as Al and Fe, greatly facilitating mineral dissolution and metal translocation within soils, thereby enhancing weathering processes.

In general, organic acids are typically highest in concentration in O horizons and decrease sharply with depth into the mineral soil (Fox and Comerford 1990; Herbert and Bertsch 1995; Richter and Markewitz 1995b). For example, in our Calhoun Experimental Forest, collections of soil water from lysimeters in soil profiles that support pine forests have soluble organic acids that decrease from about $115\,\mu\text{molc/L}$ in water that drains the O horizon to $73\,\mu\text{molc/L}$ in waters draining the A horizons, and are below detection at 60 cm and deeper (Markewitz *et al.* 1998). This decrease in organic concentrations with soil depth masks the significance of organic acids' effects on weathering in the lower soil system. Rhizospheres in lower soil systems develop in macropores, solution channels, and fracture zones and other planes of weakness (Herbert and Bertsch 1995), all of which as microenvironments that can experience relatively high concentrations of organic acids given the presence of active roots and associated microbiota.

Redox Cycling of Electron-Deficient Metals

Redox cycling in rhizospheres of relatively well aerated soils is a little-studied process with considerable potential impact on soil acidity. The presence of oxygen in relatively well-aerated soils ensures a low level of chemically reactive electrons and a preponderance of metal ions in higher valance, electron-deficient oxidation states. Contrary to the bulk soil environment with generally abundant O_2, rhizospheres can be reducing environments due to the turnover of decomposable organic compounds that are regularly added by roots (e.g., see Chapter 2). Because oxygen is actively consumed in the rhizosphere due to microbial decomposition and root respiration, steep redox gradients can develop between the near-root environment and the surrounding soil. One

TABLE 8.3 Two Symbolic Representations of Reductive Dissolution and
Oxidative Precipitation of Iron in Soils

Reductive dissolution	Oxidative precipitation
Amorphous Fe(OH)$_3$	
$Fe(OH)_{3(s)} + \frac{1}{4}CH_2O + 2H^+ \rightarrow$	$Fe^{2+} + \frac{1}{4}O_2 + 2\frac{1}{2}H_2O \rightarrow$
$Fe^{2+} + \frac{1}{4}CO_2 + 2\frac{3}{4}H_2O$	$Fe(OH)_{3(s)} + 2H^+$
Goethite (FeOOH)	
$FeOOH_{(s)} + \frac{1}{4}CH_2O + 2H^+ \rightarrow$	$Fe^{2+} + \frac{1}{4}O_2 + 1\frac{1}{2}H_2O \rightarrow$
$Fe^{2+} + \frac{1}{4}CO_2 + 1\frac{3}{4}H_2O$	$FeOOH_{(s)} + 2H^+$

visible outcome can be rhizosphere-induced mottling: that is where rhizo-spheres reduce Fe, potentially mobilizing Fe^{II} to more oxidized zones nearby. The consequences for the soil's acid budget are hypothetically enormous: approximately two moles of H^+ are produced or consumed for every mole of Fe oxidized or reduced, respectively (Table 8.3).

Such redox reactions have fascinating correlates in anaerobic soils subject to regularly high-water tables (e.g., Brinkman 1970; van Breemen 1988; Van Ranst and De Coninck 2002). In regularly anaerobic soils, roots are often the main local sources of oxygen, as many wetland plants transport O_2 to their roots to keep them alive and active. As a consequence, iron-oxidizing bacteria precipitate Fe plaque as oxidized coatings on root surfaces (Emerson *et al.* 1999; Weiss *et al.* 2003, 2004). In seasonally waterlogged soils such as paddies and other wetlands, redox cycles of reductive dissolution and oxidative precipitation are separated in time: during wet seasons and high-water tables, Fe^{III} is reduced and acidity consumed; during dry seasons, Fe^{2+} is oxidized and acidity produced. Brinkman (1970) described the consequences of these redox cycles on acidity and weathering of Pakistani wetlands and named this Fe-redox cycling "ferrolysis."

Few studies have considered how these redox processes of wetlands may be related to redox processes in generally well-aerated soils that typically experience high redox potential. Nonetheless, in relatively well-aerated soils, two processes facilitate rhizosphere-induced mottling: the ready supply of organic reductants from root and microbial activity, and the consumption of O_2 by respiration in the near-root environment. Both combine to reduce Fe^{III}: organic reductants can be adsorbed to oxide/hydroxide surfaces which facilitate Fe^{III} reduction in surface chemical reactions; if rapid rhizosphere respiration is accompanied by sluggish O_2 resupply, redox potential can poten-tially plummet. Under these conditions, Fe^{II} hypothetically enters a soluble phase and is transported out of the rhizosphere following a sequence of reac-tions outlined by Stone (1986): (1) initial adsorption and complex formation of organic reductant (QH_2) and oxide surface; (2) electron transfer; and (3) dissolution.

$$> Fe^{III}OH + QH_2 \leftrightarrow > Fe^{II}OH^- + QH + H^+ \tag{8.1}$$

$$> Fe^{III}OH + QH \leftrightarrow > Fe^{II}OH^- + Q + H^+ \tag{8.2}$$

$$> Fe^{II}OH^- \rightarrow Fe(H_2O)_6{}^{2+} \tag{8.3}$$

where the symbol $>$ represents bonding to surface metals in the oxide lattice, and QH_2, QH, and Q are hydroquinone reductant, semiquinone, and quinone, respectively. The impact of these reactions on Fe mobility is hard to overestimate. Reduction of Fe^{III} increases iron solubility with respect to oxide/hydroxide phases by as much as eight orders of magnitude (Stumm and Morgan 1996; Stumm and Sulzberger 1992).

Reductive dissolution of Fe^{III} consumes protons in the rhizosphere but upon translocation of Fe^{II} to adjacent but more oxidized microsites, Fe^{II} encounters soluble O_2 and is oxidatively precipitated. The oxidation is likely driven microbially and also produces protons which facilitate cation exchange and mineral weathering via surface chemical reactions in the bulk soil environment (Figure 8.6). Reaction kinetics of adsorbed Fe^{II} at pH <5 is relatively rapid compared to aqueous Fe^{II} (Wherli 1990), and we hypothesize that oxidation of adsorbed Fe^{II} rapidly yields co-adsorbed H^+, which protonate cation exchange and pH-dependent sites on oxides and stimulate mineral weathering. Thus, rhizosphere-induced mottling may affect great changes in acid–base status and

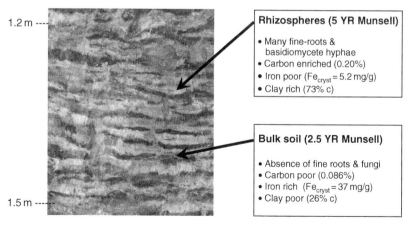

1.2 m ----

Rhizospheres (5 YR Munsell)

• Many fine-roots &
  basidiomycete hyphae
• Carbon enriched (0.20%)
• Iron poor ($Fe_{cryst} = 5.2$ mg/g)
• Clay rich (73% c)

Bulk soil (2.5 YR Munsell)

• Absence of fine roots & fungi
• Carbon poor (0.086%)
• Iron rich ($Fe_{cryst} = 37$ mg/g)
• Clay poor (26% c)

1.5 m ----

FIGURE 8.6 Pronounced rhizosphere-initiated redoximorphic features that demonstrate effects of Fe-redox cycling in B horizons at Calhoun Experimental Forest, SC. The close-up photo is of a soil excavation at 1.2–1.5 m depth. See Plate 3.

mineral weathering in soils due to steep redox gradients and spatial separation of microsites of relatively high and low redox potential.

The close correspondence of rhizospheres and soil redoximorphic features (Fimmen 2004) is observable in many soils, which supports a hypothesis that rhizosphere-stimulated Fe-redox cycling significant controls soil acid–base reactions. In humid climates, all but the most well-drained soils experience at least temporary periods of saturation during which electron-deficient Fe and Mn oxides and hydroxides can function as electron acceptors in microbially mediated reactions. In the southern Piedmont of the southeastern North America, a region nearly 20 million hectares in area, more than half of the mapped soil series have official descriptions that indicate redoximorphic features in B horizons.

Estimates of rates of Fe cycling are not well quantified, although the significance of such redox reactions to soil acidification and mineral weathering can be readily demonstrated with chemical data from Ultisols at the Calhoun Experimental Forest. In this soil's B horizon, the total content of KCl-exchangeable acidity per 1 m of B horizon is on the order of $500 \, k \, molc \, ha^{-1}$ and the $BaCl_2$–TEA exchangeable acidity (buffered at pH 8.2) is about double or triple that, to as much as $1500 \, k \, molc \, ha^{-1}$, again per 1 m of B horizon. We have previously estimated that to create 1 m of kaolinite-dominate B horizon at the Calhoun from the original bedrock of granite gneiss requires the consumption of about $100,000 \, k \, molc \, ha^{-1}$ of acid-neutralizing capacity in granite-gneiss (Richter and Markewitz 2001). In other words, to create 1 m of kaolinitic B horizon requires $100000 \, k \, molc \, ha^{-1}$ of acidity to have reacted with and cosumed the acid-neutralizing capacity of the parent geologic material. Remarkably, only about 0.5–1.5 percent of that acidity still resides on the B horizon's cation exchange sites. We can use these data to evaluate the significance of redox cycling of Fe by estimating the content of Fe^{III} that coats the surfaces of the kaolinite, quartz, and other particles in B horizons. Dithionite-citrate-bicarbonate extractions of the B horizon recovers between about 25 and $200 \, cmolc \, kg^{-1}$ of Fe which if taken to be Fe^{III} could represent 50–$400 \, cmolc \, kg^{-1}$ of H^+ generation equivalent to about 7500–$60,000 \, k \, molc \, ha^{-1}$ of the total $100,000 \, k \, molc \, ha^{-1}$ that has reacted with the parent rock to form the kaolinite-dominated soil. Rhizosphere effects on redox cycling of Fe requires much greater study with regard to its impact on mineralogy, acidification, weathering, and soil formation. These calculations reinforce the importance of conceiving of the rhizosphere broadly in space and time.

Carbonic Acid System

Respiration is a central process of ecosystems, and organic-matter decomposition and plant-root respiration elevate belowground CO_2 greatly. Soil's elevated CO_2 stimulates carbonic acid weathering with mineral surfaces and thus significant cation exchange and weathering dissolution (Reuss and Johnson 1986; Amundson and Davidson 1990; Richter and Markewitz 1995b; Oh and Richter 2004). Carbonic acid weathering involves all three phases of the soil system: CO_2 in the gas phase, carbonic acid and associated ions in the liquid phase, and in the solid phase, protons and carbonates interact with cation exchange, mineral surfaces, and mineral structures. Since partial pressures of CO_2 typically increase with soil depth, B and C horizons are subject to the main brunt of the carbonic acid system's attack. Moreover, since rhizospheres are the main sources of CO_2 in the subsoil, we expect that carbonic acid weathering is most greatly elevated in subsoil rhizospheres. Sorensen (1997) illustrated that respiration depleted O_2 in the near-root environment, we can assume this pattern is coupled with a marked increase in CO_2.

Since $H_2CO_3^*$ (the sum of dissolved and hydrated CO_2) is a very weak acid with a pK_{a1} of 6.36 (Stumm and Morgan 1996), the carbonic acid weathering system is widely conceived to be self-limiting in its effects on soil acidification and weathering (Reuss and Johnson 1986). However, $H_2CO_3^*$ can be an effective acidifying agent even at relatively low pH, as pure H_2CO_3 (hydrated CO_2) is a much stronger acid than $H_2CO_3^*$, and has even been estimated to have a pK_{a1} of about 3.8 at $25°C$ (Snoeyink and Jenkins 1980). The little appreciated, relatively strong acidity of H_2CO_3 may be a critical feature of the chemistry of soil carbonic acid, especially because CO_2 ranges commonly between 1 and 10 percent in bulk soil atmospheres. Elevated partial pressure of soil CO_2 ensures relatively high concentrations of $H_2CO_3^*$ in solution and ensures that protons of even a small fraction of $H_2CO_3^*$ will dissociate, despite low pH, due to the low pK_{a1} of pure H_2CO_3. Equilibrium calculations indicate that in situ pH is depressed from 5.65 (at atmospheric CO_2) to 4.9 and 4.4 in dilute soil waters at equilibrium with 1–10 percent CO_2, respectively (Table 8.4), and that HCO_3^- will increase from 3 to 15 and $46 \mu mol\,L^{-1}$ in dilute soil water. These values are very close to what is measured by titration

TABLE 8.4   Solution pH of Low Ionic Strength Solutions in Equilibrium with CO_2 at Different Partial Pressures. The Soil Atmosphere at >1 m Depths of Many Soils Ranges up to 5–10 percent CO_2, and in Atmospheres of Rhizospheres may Exceed 10 percent

CO_2 (%)	pH	HCO_3^- $(mmol\,L^{-1})$
0.036	5.65	0.0029
1.0	4.9	0.0145
5.0	4.6	0.036
10	4.4	0.046
100	3.9	0.145

in soil–water collections from 2 to 6 m depth in the extremely acid Ultisols of the Calhoun forest (Markewitz *et al.* 1998). Although CO_2 may rarely lower soil solution pH below 4.5 and will not rapidly mobilize much Al from soil profiles to stream and river waters (Reuss and Johnson 1986), elevated subsoil and rhizosphere CO_2 ensures that carbonic acid stimulates mineral dissolution and creates Al-saturated soils once weatherable minerals are consumed.

Two lines of evidence support this perspective of the potency of carbonic acid. A first line of evidence comes from laboratory studies (Oh and Richter 2004) in which solutions equilibrated with varying pressures of CO_2 were used to extensively leach soils that had a range of cation exchange capacities and weatherable minerals. Cation exchange was the dominant mechanism supplying cations to solution in these leaching studies which greatly diminished soil base saturation. Carbonic acid leaching displaced nearly all exchangeable base cations from two of three soils tested, and in one Ultisol, even 1 percent CO_2 displaced all exchangeable base cations and even elevated Al in soil solution.

Second, many Ultisols and Oxisols are underlain by deep saprolites or extremely acidic C horizons that represent soil conditions pushed to an extreme state of weathering (Richter and Markewitz 1995). Some of these geomorphically stable profiles are 10s of meters in depth. Given that most subsoil CO_2 originates from the rhizosphere respiration, rhizosphere processes must be recognized to affect enormous volumes of bulk soil. Of the acid-producing ecological processes that can potentially acidify such enormous volumes of C horizons, rhizosphere-initiated carbonic acid and Fe-redox cycling are likely the major candidates.

8.4   OVERVIEW OF THE RHIZOSPHERE'S WEATHERING ATTACK

Whether the perspective is one of mechanics or of chemistry, the rhizosphere represents a highly significant interface between biology and geology, an interfacial environment with broad consequences for earth's biogeochemistry and soil formation.

We started this chapter by noting that the scientific literature on the rhizosphere has historically been narrowly focused in space and time. While the focus of the rhizosphere as microsite has helped us understand how actively growing roots create special habitats for roots and microbes, rhizospheres also have much larger scale effects on soil formation and biogeochemistry.

We conclude with a summary for considering how rhizospheres drive much of the mechanical and chemical mineral weathering and the direction of soil formation. To describe this broad rhizosphere concept, we divide the soil profile into upper and lower soil systems (Brimhall *et al.* 1991;

Richter *et al.* 1995b) because rhizosphere effects differ greatly as a function of soil depth:

- an upper soil system that includes the traditional *solum*, the O, A, E, and B horizons, and
- a lower soil system that mainly includes the C horizon or saprolite.

Over millennial time scales, the upper soil system is mechanically mixed by bioturbation, a mixing that is broadly rhizospheric affected. Root pressures abrade and shatter primary particles and secondary aggregates; root balls and root plates disturb soil horizons in tip-up mounds of wind-toppled trees. Below the B horizon, however, the C horizon is more sedentary due in part to less root penetration. The mechanical mixing and pressures of growing roots accelerates rhizosphere that are chemically derived.

The chemical attack of ecosystems on soil minerals is strongly mediated by rhizospheres. In upper soil systems, rhizospheres extensively affect acidity due to root uptake of nutrient ions; in the lower system such uptake effects are highly localized within rhizospheres. Rhizosphere production of organic acids is patterned similarly to that of nutrient uptake: broadly extensive in the upper system, more spatially explicit in the lower system. Organic acids, derived from rhizodeposition and from oxidative products of decomposition, weather minerals via proton-exchange reactions, by complexing metal cations such as Al and Fe, or by serving as electron sources for the redox-cycling of electron deficient metals.

In addition to organic acids, a variety of organic compounds are added to rhizospheres by roots and microbes, many of which can facilitate reduction of metals, especially Fe and Mn, which are mobilized out of the rhizosphere only to precipitate and oxidize on contact with soluble O_2. This redox-cycling phenomenon is likely further promoted by consumption of O_2 via rhizosphere respiration. Redox-cycling of Fe and Mn affects major fluxes of protons and given the spatial separation of reduced and oxidized microsites, such acid–base dynamics may exert strong control over mineral weathering and cation exchange throughout the soil profile.

Lastly, carbonic acid, which often increases with soil depth, is likely to be greatly elevated in rhizospheres as well, given that rhizospheres are microsites of concentrated root and microbial respiration. Carbonic acid is especially important in lower soil systems and concentration gradients of CO_2 are often steep from subsoil to the soil surface and from rhizospheres to the bulk soil itself. Recent evidence suggests that the potential for carbonic acid to weather minerals and acidify even already acidic soils cannot be underestimated.

In concert, a variety of rhizosphere processes alter mineral surfaces, attack mineral structures, and over time consume weatherable soil minerals, all helping to direct soil formation. Chemical elements are released by the combined effects of mechanical and chemical weathering, taken up by plants and microbes to meet nutritional requirements, adsorbed to organo and mineral surfaces, recombined into secondary clay minerals, and leached to groundwaters, rivers, lakes, and eventually to the ocean. Over pedogenic time, on stable landforms, the ultimate soil products of such rhizosphere-assisted weathering are advanced weathering-stage soils, such as Ultisols and Oxisols. Few chemical elements are insoluble enough to resist transformations and transportation by weathering environments, due not in small part to the intense physical and chemical effects of the rhizosphere.

The concept of the rhizosphere has been significant to ecological, biological, agronomic, and forestry sciences in its first 100 years of its use. During its second century of use, the dynamics of rhizosphere response to environmental change will become a focus of intense study as the concept of rhizosphere as continuum is enriched by details of how rhizospheres interact with the whole soil profile.

ACKNOWLEDGEMENTS

We thank Dharni Vasudevan of Bowdoin College; Mark Williams, Larry T. West, and Daniel Markewitz, all of the University of Georgia; and Chuck Davey of North Carolina State University for discussions, teaching, and for the use of laboratories.

REFERENCES

Alban, D.H. 1982. Effects of nutrient accumulation by aspen, spruce, and pine on soil properties. *Soil Science Society of America Journal* 46:853–861.

Amundson, R.G., and E.A. Davison. 1990. Carbon dioxide and nitrogenous gases in the soil atmosphere. *Journal of Geochemical Exploration* 38:13–41.

Anderson, T.A., D.P. Shupack, and H. Awata. 2002. Biotic and abiotic interactions in the rhizosphere: Organic pollutants. In: Huang, P.M., Bollag, J.-M., and Senesi, N. (eds) *Interactions Between Soil Particles and Microorganisms*, John Wiley & Sons, Chichester, pp. 439–455.

April, R., and D. Keller. 1990. Mineralogy of the rhizosphere in forest soils of the eastern United States. *Biogeochemistry* 9:1–18.

Binkley, D., and D.D. Richter. 1987. Nutrient cycles and H^+ budgets of forest ecosystems. *Advances in Ecological Research* 16:1–51.

Boyle, J.R., and G.K. Voigt. 1973. Biological weathering of silicate minerals. Implications for tree nutrition and soil genesis. *Plant and Soil* 38:191–201.

Brimhall, G.H., O.A. Chadwick, C.J. Lewis, W. Compston, I.S. Williams, K.J. Danti, W.E. Power, D. Hendricks, and J. Bratt. 1991. Deformational mass transport and invasive processes in soil evolution. *Science* 255:695–702.

Brinkman, R. 1970. Ferrolysis, a hydromorphic soil forming process. *Geoderma* 3:199–206.

Buol, S.W., F.D. Hole, R.J. McCracken, and R.J. Southard. 1997. *Soil Genesis and Classification*, 4th edn. Iowa State University Press, Ames.

Chadwick, O.A., G.H. Brimhall, and D.M. Hendricks. 1990. From a black to a gray box – a mass balance interpretation of pedogenesis. *Geomorphology* 3:369–390.

Clark, F.E. 1949. Soil microorganisms and plant roots. *Advances in Agronomy* 1:241–288.

Coyne, M. 1999. *Soil Microbiology*. Delmar Publishers, Albany.

Curl, E.A., and B. Truelove. 1986. *The Rhizosphere*. Springer-Verlag, Berlin.

Denison, R.F. 2001. Ecologists and molecular biologists find common ground in the rhizosphere. *Trends in Ecology and Evolution* 16:535–536.

Dexter, A.R. 1987. Compression of soil around roots. *Plant and Soil* 97:401–406.

Driscoll, C.T., and G.E. Likens. 1982. Hydrogen ion budget of an aggrading forest ecosystem. *Tellus*. 34:283–292.

Duchaufour, P. 1982. *Pedology*. George Allen & Unwin, London.

Emerson, D., J.V. Weiss, and J.P. Megonigal. 1999. Iron-oxidizing bacteria are associated with ferric hydroxide precipitates (Fe-plaque) on the roots of wetland plants. *Applied and Environmental Microbiology* 65:2758–2761.

Fimmen, R.A. 2004. Organic geochemical processes in the South Carolina Piedmont. PhD dissertation, Duke University, Durham.

Fox, T.R., and N.B. Comerford. 1990. Low-molecular-weight organic acids in selected forest soils of the southeastern USA. *Soil Science Society of America Proceedings* 54:1139–1144.

Herbert, B.E., and P.M. Bertsch. 1995. Characterization of dissolved and colloidal organic matter in soil solution: A review. In: McFee, W., and Kelly, J.M. (eds) *Carbon Forms and Functions in Forest Soils*. Soil Science Society of America, Madison, pp. 63–88.

Hiltner, L. 1904. Über neuere Erfahrungen und Probleme auf dem Gebiet der Bodenbakteriologie und unter besonderer Berücksichtigung der Gründüngung und Brache. *Arbeiten der Deutschen Landwirtschafts Gesellschaft* 98:59–78.

Jackson, M.L., and G.D. Sherman. 1953. Chemical weathering of minerals in soils. *Advances in Agronomy* 5:221–317.

Johnson, D.W., and S.E. Lindberg. 1992. *Atmospheric Deposition and Nutrient Cycling in Forest Ecosystems*. Springer-Verlag, New York.

Katznelson, H. 1946. The rhizosphere effect of mangels on certain groups of micro-organisms. *Soil Science* 62:343.

Lapeyrie, F., G.A. Chilvers, and C.A. Bhem. 1987. Oxalic acid synthesis by the mycorrhizal fungus *Paxillus involutus* (Battsch. ex. Fr.) Fr. *New Phytologist* 106:139–146.

Lovejoy, A.O. 1936. *The Great Chain of Being*. Harvard University Press, Cambridge.

Lynch, J. 1990. *The Rhizosphere*. John Wiley & Sons, Chichester.

Markewitz, D., D.D. Richter, H.L. Allen, and J.B. Urrego. 1998. Three decades of observed soil acidification at the Calhoun experimental forest: Has acid rain made a difference? *Soil Science Society of America Journal* 62:1428–1439.

Misra, R.K., A.R. Dexter, and A.M. Alston. 1987. Maximum axial and radial growth pressures of plant roots. *Plant and Soil* 95:315–326.

Oh, N.H., and D.D. Richter. 2004. Soil acidification induced by elevated atmospheric CO_2. *Global Change Biology* 10:1936–1946.

Pierre, W.H., J. Meisinger, and J.R. Birches. 1970. Cation-anion balance in crops as a factor in determining the effects of nitrogen fertilizers on soil acidity. *Agronomy Journal* 62:106–112.

Qualls, R.G., and B.L. Haines. 1991. Geochemistry of dissolved organic nutrients in water percolating through a forest ecosystem. *Soil Science Society of America Journal* 55:1112–1123.

Reuss, J.O., and D.W. Johnson. 1986. *Acid Deposition and the Acidification of Soils*. Springer-Verlag, New York, USA.

Richter, D.D. 1986. Sources of acidity in some forested Udults. *Soil Science Society of America Journal* 50:1584–1589.

Richter, D.D., and L.I. Babbar. 1991. Soil diversity in the tropics. *Advances in Ecological Research* 21:316–389.

Richter, D.D., and D. Markewitz. 1995a. How deep is soil? *BioScience* 45:600–609.

Richter, D.D., and D. Markewitz. 2001. *Understanding Soil Change: Soil Sustainability Over Millennia, Centuries, and Decades.* University of Cambridge, United Kingdom.

Richter, D.D, B. Browne, K.H. Dai, T. Huekel, D. Markewitz, H. Stevens, and A. Stuanes. (submitted) Soil compaction by the tap root of a growing tree. *Soil Science Society of America Journal* (submitted)

Richter, D.D., D. Markewitz, C.G. Wells, H.L. Allen, J. Dunscomb, K. Harrison, P.R. Heine, A. Stuanes, B. Urrego, and G. Bonani. 1995b. Carbon cycling in an old-field pine forest: Implications for the missing carbon sink and for the concept of soil. In: McFee, W., and Kelly, J.M. (eds) *Carbon Forms and Functions in Forest Soils.* Soil Science Society of America, Madison, pp. 233–251.

Rovira, A.D., and C. Davey 1974. Biology of the rhizosphere. In: Carson, E.W. (ed.) *The Plant Root and its Environment.* University Press of Virginia, Charlottesville, pp. 153–204.

Snoeyink, V.L., and D. Jenkins. 1980. *Water Chemistry.* Wiley, New York, 463 pp.

Sollins, P., C.C. Grier, F.M. McCorison, K. Cromack, R. Fogel, and R. L. Fredriksen. 1980. The internal element cycles of an old-growth Douglas-fir ecosystem in western Oregon. *Ecological Monographs* 50:261–285.

Sorensen, J. 1997. The rhizosphere as a habitat for soil microorganisms. In: Van Elsas, J.D. *et al.* (eds) *Modern Soil Microbiology.* Marcel Dekker, Inc., New York, USA.

Stone, A.T. 1986. Adsorption of organic reductants and subsequent electron transfer on metal oxide surfaces. In: Davis, J.A., and Hayes, K.F. (eds) *Geochemical Processes at Mineral Surfaces.* American Chemical Society, Chicago, pp. 446–461.

Stumm, W., and J.J. Morgan. 1996. *Aquatic Chemistry.* 3rd ed. Wiley-Interscience, New York, USA.

Stumm, W., and B Sulzberger. 1992. The cycling of iron in natural environments: Considerations based on laboratory studies of heterogeneous redox processes. *Geochimica et Cosmochimica Acta* 56:3233–3257.

van Breemen, N. 1988. Effects of seasonal redox processes involving iron on the chemistry of periodically reduced soils. In: Stucki, J.W., Goodman, B.A., and Schwertmann, U. (eds) *Iron in Soils and Clay Minerals D.* Reidel Publishing Company, Dordrecht, pp. 797–810.

van Breemen, N., P.A. Burrough, E.J. Velthorst, *et al.* 1982. Acidification from atmospheric ammonium sulphate in forest canopy throughfall. *Nature* 299:548–550.

Van Ranst, E., and De Coninck, F. 2002. Evaluation of ferrolysis in soil formation. *European Journal of Soil Science* 53:513–519.

Weiss, J.V., D. Emerson, S.M. Backer, and J.P. Megonigal. 2003. Enumeration of Fe(II)-oxidizing and Fe(III)-reducing bacteria in the root zone of wetland plants: Implications for a rhizosphere iron cycle. *Biogeochemistry* 64:77–96.

Weiss, J.V., D. Emerson, and J.P. Megonigal. 2004. Geochemical control of microbial Fe(III) reduction potential in wetlands: Comparison of the rhizosphere to non-rhizosphere soil. *FEMS Microbiology Ecology* 48:89–100.

Wherli, B. 1990. Redox reactions of metal ions at mineral surfaces. In: Stumm, W. (ed.) *Aquatic Chemical Kinetics: Reaction Rates of Processes in Natural Waters*, John Wiley & Sons, New York, pp. 311–336.

INDEX

Page numbers referring to figures are followed by 'f' and those referring to table are followed by 't'

Printed and bound by CPI Group (UK) Ltd, Croydon, CR0 4YY

03/10/2024

01040412-0002